Winning with Operational Excellence

A Practical Manual for Managers and Engineers for Implementing
Lean Manufacturing and the Toyota Production System

Winning with Operational Excellence

A Practical Manual for Managers and Engineers for Implementing Lean Manufacturing and the Toyota Production System

Sarv Singh Soin

Copyright © 2012, Sarv Singh Soin

All rights reserved. No part of this book may be reproduced, stored, or transmitted by any means—whether auditory, graphic, mechanical, or electronic—without written permission of both publisher and author, except in the case of brief excerpts used in critical articles and reviews. Unauthorized reproduction of any part of this work is illegal and is punishable by law.

ISBN 978-1-105-32267-9

"This is a book about manufacturing management: it's not a book about manufacturing theories – it's written by a practitioner for practitioners. It is an excellent quick and practical reference of leading practices of excellence in manufacturing. In other words, for building and maintaining resilient manufacturing operations. I challenge any manufacturing manager in the world to know and apply the practices in this book by heart - chapter by chapter."

Dr. Dieter Legat, Co-founder, The Delta Institute, Switzerland

"In this example packed book Soin shares his profound knowledge of the methodologies that will enable executives and managers to achieve operational excellence. He educates the practitioner in enough detail to get the most out of the tools and techniques in their daily work. On top of this he explains the sequence of adoption and what to look for as signs of progress or opportunities along the way. The path to operational excellence is challenging, but with this book as a companion the journey became a lot easier and the chances of success higher."

Vice President Quality System Deployment, Philips Healthcare
Formerly Vice President Business Excellence, Philips Healthcare

"Consistent and sustainable Operational Excellence (OE) is the holy-grail for any manufacturing organization. Ask any experienced manufacturing professional and he or she will tell you OE remains the cornerstone of total customer satisfaction. True time tested and industry leading OE is elusive even to the standard-bearers, as was evidenced by the recent crisis at Toyota. However, OE is not a destination, and **Winning with Operational Excellence** *does a remarkable job of reminding the reader of the perpetual nature of improvement and the unending journey an organization must take in the pursuit of perfection.*

In unpretentious and simple "factory language", Soin distils the key techniques that have been proven successful over the years, and provides the manufacturing professional with a Swiss Army Knife of solutions, any of which can be deployed readily and easily.

This is a book for all levels. For the beginner or the learner, Soin practically hands out solutions on a plate. Apply the prescribed approaches and you will be solving problems and making lasting improvements at your workplace. Your eye for improvement will be sharpened, and you will uncover opportunities previously unseen."

Dharma Nadarajah, Senior Vice President, Venture Corporation

Purpose

The objective of this book is to help you achieve operational excellence in manufacturing via the tools of Lean Manufacturing and the Toyota Production System. With these concepts, tools, and methodologies you will be able to increase the wealth of your corporation and create higher quality and more value for your customers. This book is for engineers and managers; it shows you how to:

- Organize your factory floor via the 5S system.
- Get high performance via standard work.
- Manage and improve production cycle time for higher productivity.
- Implement JIT (just-in-time) manufacturing via Kanban system.
- Manage machines with SMED, quick setup, and TPM (Total Productive Maintenance).
- Achieve breakthrough goals with the PDCA improvement cycle.
- Achieve the highest quality, a strong Kaizen effort, Poka Yoke, and aim for zero defects.
- Put all the tools together to get continuous flow manufacturing via cells or linear lines.
- Develop employees and suppliers.
- Audit your strengths and weaknesses in lean manufacturing.
- Conduct factory floor or system wide Value Stream analysis to streamline your operations.
- Do the right things for the company to achieve success via Hoshin Kanri planning; it will also enable you to use TOC (Theory of Constraints) to prioritize your planning.
- Get started, improve, and win with operational excellence.

Author's Preface

I have been deeply satisfied with the success of my original book, *Total Quality Essentials*. I have received numerous requests to go beyond Total Quality Management, apply quality thinking to my manufacturing experience, and write a book on excellence in running a manufacturing operation.

The objective of this book is to help you achieve operational excellence in manufacturing via the tools of Lean Manufacturing and the Toyota Production System. With these concepts, tools, and methodologies you will be able to increase the wealth of your corporation and create higher quality and more value for your customers. Hence the title: *Winning with Operational Excellence*.

Many of the tools and methodologies of operational excellence are derived from Lean Manufacturing, the Toyota production System, and Total Quality Management. However, according to the *Shingo Prize for Operational Excellence*, there has been a tendency to disassemble the various tools and methodologies into sub-sets, and promote them as standalone tools. For example: JIT (Just-in-time) manufacturing, continuous flow, TPM (Total Productive Maintenance), A3 Improvement process, and Lean Six-Sigma. This has resulted in a haphazard tool-driven attempt to copy these concepts and has delayed understanding and optimization of manufacturing efficiency. As a result, modern management, with its short-term focus on fads, has prevented manufacturing from fully benefitting from this wealth of knowledge. Our purpose in this book is to put all the relevant tools and methodologies together in a unified lean manufacturing system

This book helps you assess your strengths and weaknesses, identifies opportunities, introduces you to many of the successful tools and methodologies, puts the tools together in a systematic way, and integrates everything into a unified lean manufacturing system. It will help you improve efficiency, productivity, and quality in the factory and lay a strong foundation for generating a stream of attractive products and services.

This book is for engineers and managers. The engineer can use it to drive specific improvements on the line. The manager should use it first to assess the overall health of the system, and then to set priorities via the appropriate planning process. Together, the engineer and manager can move towards a unified lean manufacturing system to incorporate all the methods discussed here, and *Win with Operational Excellence*.

Author Information

Soin started work as an engineer in Hewlett-Packard Company. His twenty years of work experience at HP includes assignments in computer and printer manufacturing, quality management, supply chain management, and General Manager. Currently he is the Director of Global Quality at Venture Corporation based in Singapore. He has extensive hands-on experience in lean manufacturing and TQM, and has a Doctorate in Business Management (Supply Chain). Contact the author at LinkedIn or the website: www.opex-usa.org

Acknowledgements and Dedication

I am extremely indebted to the numerous people who have contributed or guided me in my effort to write this book: Craig Walter for his initial encouragement in starting me on this journey; Dr. Noriaki Kano for his valuable insights on quality and productivity; Dieter Legat for his observations on management's role in running a business with the highest quality; John Hamilton for his valuable inputs on Hoshin Kanri planning, edits, and the format of this book. Other contributors have been Katsu Yoshimoto, John Mustaffa, Jeremiah Josey, Kek Kiong Tan, Thomas Lee, and many others who have shared their experiences with me. Lastly, my thanks to Mona for her infinite patience and encouragement.

This book is dedicated to all engineers, professionals, and managers who strive for operational excellence. Feedback and recommendations for improvement are most welcome.

Table of Contents

Author's Preface ix

       Acknowledgements and Dedication xi

Chapter 1   **Operational Excellence** 1

    What Are Our Expectations For Operational Excellence? 1

       Toyota Production System 2

       Lean Manufacturing System 3

    Principles and Best Practices of Operational Excellence 4

       Process Thinking 4

       Scientific Management 4

       Quality First 4

       Eliminate Waste 4

       Continuous Flow 5

       Employee Participation and Development 5

       Constancy of Purpose 5

       Create Value for the Customer 5

    What We Will Cover in This Text 6

    Summary: Operational Excellence in the Factory 7

Chapter 2   **The Basics: Getting Organized** 9

    Overview 9

    The 5S System 9

    Objectives of the 5S System 10

    First Step: *Seiri* or Sort and Organize 10

       The Red-tag Process 12

    Second Step: *Seiton,* Stabilize, or Set in Order 13

       Who's Minding the Kitchen? 14

    Third Step: *Seiso* or Shine 15

	Fourth Step: *Seiketsu* or Standardize	16
	Fifth Step: *Shitsuke* or Sustain	17
	Sustaining 5S in the Factory	18
	Conducting 5S audits	19
	Tracking 5S Progress	22
	Barriers to a Good 5S System	23
	Do Customers Really Expect to See an Organized Factory?	26
	Summary: The Basics – Getting Organized	26
Chapter 3	**High Performance: Standard Work**	**27**
	Overview	27
	Preparing and Implementing Standard Work	27
	Content of a Standard Work	28
	Sample Standard Work	29
	Preparing Standard Work	31
	Improving Standard Work	35
	Genchi Genbutsu: The Gemba Walk	37
	The Ohno Circle	40
	The Training Process	43
	Summary: Standard Work	44
Chapter 4	**Efficiency: Manage Cycle Time**	**45**
	Overview	45
	Takt Time, Cycle Time, and Total Cycle Time	45
	Why Manage the Factory Floor via a Takt Time System?	47
	Takt Time Vs Cycle Time	47
	Challenges with the Takt Time and Cycle Time System	49
	Line Balancing	51
	Project: Line Balancing	52
	Yamazumi Charts	57
	Managing Bottlenecks Due to Material and Quality Hiccups	59
	Workstation Cycle Time Vs Overall Manufacturing Cycle Time	59
	The Impact of Variability on Production Efficiency	61

	Leveled Production 61

          Leveled Production Vs Build to Order 62

          Rebalancing Lines after Change in Demand or Forecast 62

    Summary: Efficiency: Manage Cycle Time 63

Chapter 5   **Just-In-Time Production: Kanban System** **65**

    Overview 65

    The Kanban System 66

        Description of the Kanban Pull System 66

    Kanban System Rules 69

        One-card Kanban and Card-less Kanban Systems 71

        Electronic kanban 72

    Getting Started With a Kanban System within the Factory 74

        How Many Kanban Cards? 75

    Implementing a Kanban System with Suppliers 80

    Major Challenge with JIT and Kanban System 81

    Summary: Just-In-Time Production and Kanban 82

Chapter 6   **Machine Management: Quick Setup and TPM** **83**

    Overview 83

    Reducing Production Lead-time 83

        How long does it take you to Change a Tire and Refuel? 84

    Quick Setup of Machines and Equipment 85

        Project: Changeover and Setup Improvement 87

    Specific Benefits of Quick Setup Time 89

    Total Productive Maintenance (TPM) 90

    Typical Pattern of Machine Failure 90

    Causes and Prevention of Machine Failures 91

        What is The Failure Goal for Aircraft? 92

    Measuring Machine Failure Rate and Productivity 93

    Basic Requirement for TPM 93

    Summary: Machine Management: Quick Setup & TPM 101

Chapter 7 **Quality: Improvement and Control** 103

    Overview 103

    Quality and Customer Satisfaction 104
        Quality Definition *104*

    Customer Satisfaction 105
        The Kano Model: Two Dimensions of Quality *105*
        Reactive activities *106*
        Proactive activities *106*

    Quality Improvement and the PDCA Cycle 107
        Problem Solving Styles *107*

    The PDCA Cycle 107

    Detailed PDCA cycle 109
        Project: Improvements Using PDCA Methodology *116*

    Benefits of the PDCA Improvement Cycle and Why It Works 122

    Uses of the PDCA Cycle 123

    Seven Quality Control Tools and Other Methodologies 124

    Types of Improvements Projects 125
        Rate of Improvement and Setting Targets *126*

    Management Improvements via the A3/A4 Improvement Process 127
        A Case Study using the A3/A4 Improvement Process *128*

    Problem-Solving Hierarchy 132

    Relationship between Improvement and Control 132

    Quality Control and Management 134
        The Need for Process and Quality Control *134*

    Statistical Process Control 135
        Types of Control Charts and Rules *136*

    Monitoring and Controlling Quality on the Line via Audits 137

    Aiming for Zero Defects 140

    Source or Incoming Quality Inspection 141
        The 10X Rule *141*

    Methodologies for Early Detection of Defects 142

Preventing Human Errors through Poka Yoke 144
Poka Yoke 145
    The Next Process Is My Customer 147
    Successive Inspection System 147
Jidoka and Andon Systems 147
Setting Quality Goals 150
Total Quality Management (TQM) 151
Objective of Total Quality Management 151
Summary: Quality 152

Chapter 8   **Continuous Flow: One-Piece Production** 155
Overview 155
One-piece Production 155
Benefits of One-piece Production 157
Challenges with One-piece Production 158
    Little's Law: Reducing Queues In the Factory 160
Little's Law and Continuous Flow Manufacturing 161
Alternative Production Systems and Layouts 161
Linear and Cellular Production 164
    Choosing between Batch, One-piece Cell, or Linear Production ... 164
Design of Linear Production Line 165
Cellular Production 167
Major Benefit of Cellular Production 168
Design of Cellular Production 168
Cellular Production Layouts 169
    Types of Cell Flows 172
    Maximizing Workflow Productivity 173
Converting from Batch to Continuous Flow Manufacturing 173
    Project: Converting from Batch to Continuous Flow 174
Minimizing Variability and Other Barriers to Continuous Flow 174
Summary: Continuous Flow and One-Piece Production 177

Chapter 9   Employee & Partner Participation and Development 179

- Overview 179
- Kaizen and Kaizen Team Activity 180
- Employee Suggestion Scheme 183
  - *The Foundation for a Suggestion Scheme is Twofold* 184
- Guidelines for a Suggestion Scheme 184
  - *What Type of Suggestions Can You Expect?* 188
- Education and Training 189
- Working with Partners and Suppliers 190
- Summary: Employee & Partner Participation and Development 190

Chapter 10   Doing the Right Things - Hoshin Kanri Planning 193

- Overview 193
  - *Making the Future Happen via Planning, Execution, & Control* 193
- Essentials of a Planning Process 194
- Hoshin Kanri Planning 194
- The Hoshin Kanri Planning Process 194
  - *Daily Management Plan* 195
  - *Launching the Plan* 195
  - *Hoshin Plan Reviews* 197
  - *Illustrating Hoshin Planning* 197
  - *Hoshin Plan: Ensuring Success* 198
  - *Relation between Hoshin Plan and Daily Management Plan* 199
- Formats and Guidelines 199
- Annual Hoshin Plan 199
  - *Objective* 200
  - *An Example of an Issue List from the Previous Year* 203
  - *Chief Executive's Objectives* 205
  - *Strategies and Performance Measures* 205
  - *Deployment and Cascading of an Objective* 208
- Hoshin Plan Deployment Matrices 209
- Alternative Hoshin Planning Format and Deployment Matrix 210
- Implementation or Project Plan 213

	Daily Management Plan	215
	Elements in a Daily Management Plan:	*215*
	Guidelines for Daily Management Plan	*215*
	Displaying Daily Management plan	*217*
	Managing Abnormalities and Deviations	*217*
	Setting Numerical Targets	217
	The Review for Hoshin Plans and Daily Management Plans	218
	Basic Methods of Analysis for Conducting a Review	*218*
	Guidelines for Conducting a Review	*219*
	Putting All the Plans Together	222
	Planning and Budgeting	*222*
	Planning and Time Management	*223*
	Comments and Criticism of Hoshin Kanri Planning	*224*
	The Theory of Constraints	225
	Using TOC for Business Planning	225
	Using the Lessons Learnt from TOC in Hoshin planning	*228*
	Leadership in the Company or Organization	229
	Summary and Conclusion	230
Chapter 11	**Value Stream Mapping**	**231**
	Overview	231
	Value Stream Mapping	231
	Value Stream Mapping Procedure	232
	Value Stream Mapping Project: Order Fulfillment	*235*
	Value Stream Mapping Project: Product Manufacturing	*237*
	Comments and Weakness of Value Stream Mapping	238
	Summary	238
Chapter 12	**Putting it All Together: Accomplish the Transformation**	**239**
	Overview	239
	Getting Started: Some Precursors	239
	Understand Your Current Situation and Where You Are Going	*239*
	Why Drive Toward Operational Excellence?	*240*
	Conduct an Overall Assessment	*240*

Assessment Checklist	*241*
Educate and Train	242
Appoint a Facilitator or Sensei	242
The Project Management Approach	242
Select an Operation in Crisis or One Needing Improvement	*243*
Appoint a Project Team and Lead	*243*
Revisit Safety Issues and the Organized Workplace	*243*
Work on the Most Critical Issue	*243*
PDCA and Project Management	*244*
The Business Plan Approach: Using Hoshin Planning	247
Challenges and Solutions for Success	248
Summary	250

Appendices **253**
Appendix 1: Glossary of Terms 253
Appendix 2: Assessing Operational Excellence 257
Appendix 3: References 261
Appendix 4: Resources and Websites 263
Index 265
Notes and References 269

Chapter 1
Operational Excellence

Excellence is a better teacher than mediocrity.
The lessons of the ordinary are everywhere.
Truly profound and original insights are to be found only in studying the exemplary.
Warren Bennis

What Are Our Expectations For Operational Excellence?

When we walk into any factory, what do we look for?

We look for a tidy and organized factory floor; we would like to see un-clogged aisle ways, production operations running efficiently in a continuous flow pattern, with little inventory, no defective products; and workers focused on their jobs. Overall, we expect an operation running smoothly like clockwork and synchronized like an orchestra.

Actually, the trained eye expects more and can determine very quickly if a factory is managed well or otherwise. There are several things we would look for:

The organized work environment: How is space utilized? The best factories use space efficiently with no waste and yet ensure safety, cleanliness, and smooth workflow. During a tour, we look at how production material is stored and moves on the factory floor. We also expect an organized and efficient layout on the production floor and warehouse.

The factory has a visual management system: Signage and directions are visible for visitors and employees – not only for direction, but also for production activities and inventory. These visual aids indicate good organization and improve overall productivity.

Efficient workforce: Employees are focused on their jobs: little chit chatting, or workers standing idle; very little worker movement. Work instructions or standards are displayed near the worker, either via posted documents or on an electronic display. An efficient factory will certify workers in multi-tasking; such certification is usually displayed on employee name badges and the production line. However, when we talk of efficient workers our expectations are: skilled and knowledgeable workers, who think and do quality work, and are treated respectfully by management. This is usually obvious after a short discussion with them. We also expect the factory staff to participate in product and quality improvements.

Smooth flowing production lines or work-cells: Product is flowing continuously and smoothly down the line; production is pulled by customer demand. There is little visible inventory or defective parts or products lying around. Similarly, if the factory runs cell or small batch manufacturing, there is little congestion and inventory.

Overall inventory levels are low: Inventory is managed via Kanban processes and work-in-process (WIP) inventory is low. A well managed Kanban process will result in little inventory, but during a tour continuous material delivery is often visible. A walk through the stockroom will show low inventory levels regardless of production volumes: *Low inventory is one of the best indicators of a factory's efficiency and leanness.*

Equipment and tools are clean and well maintained: Equipment and the equipment area are clean, with no dust, debris, or spills; maintenance schedule and checklists are accessible, updated, and displayed. Tools are placed in an orderly fashion and easy to locate and identify.

Commitment to quality is evident: Product quality is excellent – there is minimal quantity of defects in production or in backrooms. Product assembly and test yields are very good and the data is displayed in production areas.

The factory's business objectives and metrics are displayed. Worker commitment to quality is evident via improvement project or kaizen display boards. Metrics displayed include internal and outgoing quality levels, with little rework. Most important, customer satisfaction information is displayed showing very good customer feedback.

Finished goods are shipping daily and inventory is low: Production is based on "pull" requests from customers and shipping daily; inventory on finished product is low or committed to ship soon.

The above attributes[1] are a shortlist of what we expect to see in a factory that has operational excellence. But what are the methodologies to get us there?

Many of the tools and methodologies of operational excellence are derived from TPS (Toyota Production System) and TQM (Total Quality Management). However, according to the *Shingo Prize for Operational Excellence*[2], there has been a tendency to disassemble the various tools and methodologies into sub-sets which are promoted as standalone tools. For example: JIT (Just-in-time) manufacturing, Lean Six-Sigma, TQM, and TPM (Total Productive Maintenance). This has resulted in a haphazard tool-driven attempt to copy these concepts. Our purpose in this book is to put all the relevant tools and methodologies together in a unified Lean Manufacturing System

What does a company that excels in operational excellence, like Toyota, do in its quest for perfection? The answer comes from two landmark studies on the Toyota Production System and Lean Manufacturing:

Toyota Production System

Jeffrey Liker[3] of the University of Michigan, puts forward a "4 P" model (Fig 1-1) of the Toyota Production System. This lists the following: Philosophy, Process, People and Partners, and Problem Solving. Unfortunately, according to Liker, most companies put most of their energy in implementing the P or Process step. Why? We believe they do so because this is the easiest step to visualize and execute; a company can do this step for eternity and never run out of ideas. Unfortunately this will result in a shop floor improvement focus. Therefore all the 4 "Ps" are essential.

Lean Manufacturing System

Womack and Jones[4] seem to have invented the term Lean Manufacturing in their work entitled, *"Lean Thinking: Banish Waste and Create Wealth in Your Corporation."* This is a thoughtful description of the value-based business system used at Toyota Motors. The essence of their model is that all businesses must define the "value" that they produce for their customers. They must understand this "value stream" so that they can create value and reduce *muda,* or waste, and work towards perfection. Womack and Jones put forward the principles of lean manufacturing that can reduce waste and costs, reduce lead times, and improve quality and resource utilization. Strategies to achieve these principles include building products based on customer demand or pull and continuous flow manufacturing.

Next we give a list of the principles and best practices that help achieve operational excellence. These come from the principles put forward by Liker (on the Toyota Production System), Womack and Jones (on Lean Manufacturing), the *Shingo Prize for Operational Excellence*, and our own experience.

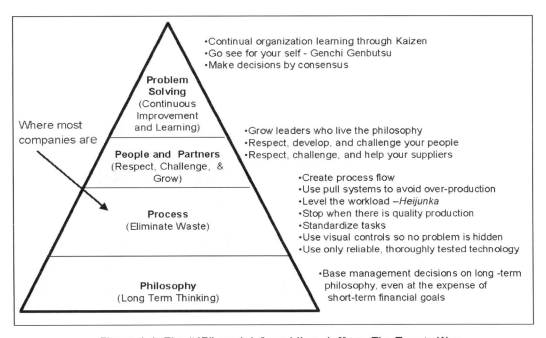

Figure 1-1: The "4P" model, from Liker, Jeffrey, *The Toyota Way.*

Principles and Best Practices of Operational Excellence

Process Thinking

Every activity is part of a process. This applies to the manufacturing and service sectors. Every organization will have a multitude of processes - some managed by individuals and some managed cross-functionally.

Why manage processes? Achieving good results is paramount in any company, because they indicate commendable performance, which is what management strives to do. Good results, however, are a *lagging indicator* of performance; only when results are achieved do we know that we have performed well. But we must be able to predict results - for this we need a *leading indicator.* A well managed process, monitored with properly selected performance metrics, can be a leading indicator for predictable results. Another important point about processes: *People, and managers, come and go (in any organization) but processes stay.* Hence there must be a system to ensure processes are well documented, and managed; otherwise there will be disruption whenever operators and staff leave.

Scientific Management

It is easier to manage and improve a process if it is well defined and documented. One of the most important tools for managing and improving products and processes is the PDCA (plan, do, check, and act) cycle. My friend and mentor, Dr. Kano, says using the PDCA cycle is *scientific management*; that is managing with facts and data instead of intuition. The PDCA cycle is used for problem solving, continuous improvement, project management, planning, and promotes learning.

Quality First

A company needs to aim for zero defects by reducing defects early in the process at suppliers and in manufacturing. This will result in less waste and defects and lead to higher quality and increased customer satisfaction and loyalty. To this end, the *quality first* policy needs to be adopted. This will lead a company towards a philosophy of high quality and continuous improvement.

Eliminate Waste

After he founded Ford Motor Company, Henry Ford[5] proposed continuous *material flow in manufacturing, standardized processes, and eliminating waste*; his factories strived to reach these goals. Toyota Motor Company and its mangers, including Taiichi Ohno, took up the challenge and identified all types of waste that had to be eliminated from the factory. Apparently, Mr. Ohno was often ruthless in his desire to drive out waste from the Toyota system, which may explain why he succeeded. Eliminating waste or *muda (*the Japanese term for waste) is one of the foundations of the Toyota Production System (TPS). Waste can be segmented into the seven wastes, shown and discussed in Table 1-1. Reducing such waste will result in a more efficient and productive factory.

However, completely eliminating these wastes would be utopia, and is not feasible. Nevertheless, our goal is to reduce continuously these wastes in the factory. This does not imply expensive automation of the entire factory; rather we desire simple but effective systems, good processes, continuous improvements of current systems, and a focused workforce.

Continuous Flow

The heart of operational excellence is continuous flow in the production process. Continuous flow of product from incoming parts from suppliers all the way to the customer can only happen with smooth and efficient processes, low defects, low WIP (work in process) inventory, predictable output, and on-time delivery of products. The tools to achieve this are basically the entire toolbox of operational excellence, which we will discuss in this text.

Employee Participation and Development

Employee participation, engagement, and involvement is a priceless commodity; it is essential for a company's success. This includes: skills-training to improve productivity, participation in improvement activity, and suggestion schemes.

Constancy of Purpose

Results cannot be achieved over the short term. Therefore, the management team must set a vision and direction and work towards achieving them. This requires *"constancy of purpose"*, per Dr. Deming, over the long term.

Create Value for the Customer

With operational excellence, there will be improvements in processes, responsiveness, and quality; furthermore the overall cost structure will be lower. These gains will allow the company to increase profits and add value for the customer in product and service features, without increasing pricing, thus making the final product more attractive and competitive. Both customers and stakeholders will benefit. This is Womack and Jones' *"create wealth in your corporation."*

Table 1-1: The Seven Wastes

1	**Overproduction**	Producing more than a factory needs is an unforgivable sin. A factory should only build what is on order and no more; building beyond requirements is sheer waste. Hence, one of the factory's objectives should be just in time (JIT) production.
2	**Unnecessary transportation**	There should be no wasted effort to move parts, materials, or finished goods into or out of an area, or between processes. That is, move material only when required. For example, the process and layout should be configured to move product to the next operation, rather having workers do the moving.
3	**Unnecessary motion**	There should be no wasted motion to pick up parts or to stack parts. To help reduce waste here, there should be fixtures, equipment, and tools that can speed up the process and reduce motion of workers.
4	**Waiting time**	There should be no wasted time waiting for work or products to arrive at a work station; nor should there be wasted time waiting for tools, parts, operating supplies, or equipment downtime. The production line should be well balanced to minimize waiting time for workers.
5	**Over-processing**	The production line should not be doing more work than is necessary. Often, many tasks can be combined or eliminated. There should be no rework or repackaging.
6	**Excess inventory**	There should be a minimum of parts waiting to be processed, and there should be little or no inventory between work stations or processes. Overall factory inventory levels should be low.
7	**Defects and poor quality**	Defect levels at assembly, test, or inspection should be very low. Hence there should be very few parts or product awaiting rework.

What We Will Cover in This Text

In this text we will cover topics that support the principles and best practices of operational excellence. There has been much focus on the tools or the one right way (the Toyota way); however, there are other successful companies that deviate from the exact model but follow the right principles, hence we will cover alternative methodologies as well.

Tools and methodologies: We will review and discuss how to implement the tools of the Toyota Production System and Lean Manufacturing. These tools will help reduce waste, improve quality, lower costs, and increase productivity. The topics include:

- The basics of getting organized – the 5S system.
- High performance via standard work.
- Managing cycle time.
- JIT (just-in-time) manufacturing via the Kanban system.
- Machine management and TPM (Total Productive Maintenance).
- How to achieve the highest quality, a strong Kaizen effort, Poka Yoke Systems, and aim for zero defects in manufacturing.
- Putting all the tools together to get continuous flow manufacturing via cells or high volume lines.
- Employee development and participation.
- How to audit your strengths and weaknesses in lean manufacturing.
- How to conduct factory floor or system wide Value Stream analysis to streamline your operations.
- Doing the right things to achieve success via Hoshin Kanri planning; we also discuss TOC (Theory of Constraints) and show how it can be used to prioritize your planning.
- Getting started or improving your current effort.

Summary: Operational Excellence in the Factory

We have discussed the principles and practices of operational excellence and the topics we will cover. The benefits of operational excellence include higher productivity, improved quality and reliability of products and services, lower costs, lower inventories and floor space with resulting better cash flow, increased customer satisfaction and value for the customer; all this leads to increased market share and profits.

In summary, operational excellence is:

A comprehensive effort of waste reduction, continuous improvement, and manufacturing excellence by all resources in the company with the goal of increasing wealth in your corporation and providing the highest quality and value to your customers.

Onward!

Chapter 2
The Basics: Getting Organized

"The shop floor is a reflection of management."
Taiichi Ohno

Overview

When we walk into a factory, the first impression is critical – does the place look clean and organized? How is the housekeeping? Is the production floor full of activity; are the employees properly attired, orderly, and organized? We believe that how the shop floor looks is a reflection on how management views and manages the factory and the entire business. In fact, management's attitude to housekeeping and orderliness tells much more than the financial statement.

In this chapter we will discuss how to get the shop floor and factory organized by using the 5S System; we will discuss how to manage, maintain, and sustain such a system; and most important, we will discuss how the 5S system can set the stage for advanced operational excellence activities.

The 5S System

Cleanliness and orderliness in a factory were proposed in 1911 by Frederick Taylor[6] in his text on *Scientific Management*. Henry Ford[7] picked up on this when he started the Ford Motor Company, he wrote: *"Good work is difficult excepting with good tools used in clean surrounding."* Indeed, Ford started a CANDO (Cleaning up, Arranging, Neatness, Discipline, and Ongoing Improvement) process, which predates the 5S system, but clearly sets the stage for its adoption and improvement by Japanese companies. Incidentally, another innovation that the Japanese companies picked up from Ford[8] was its suggestion system, which is discussed later. At we write this section, it looks like Ford is going back to its roots and having a well-deserved resurgence.

More recently, Hiroyuki Hirano[9] in Japan popularized the 5S system of workplace organization. The 5S system refers to 5 activities beginning with the letter 'S", namely Seri (or Sort), Seiton (or Stabilize), Seiso (or Shine), Seiketsu (or Standardize), and

Shitsuke (Sustain). Some operations or factories that practice 5S have actually added a sixth item – safety – thus calling it the 6S system.

Objectives of the 5S System

Many operations that use the 5S system consider it a housekeeping program. But it's much more than that. The 5S system has several objectives: It lays the foundation for good manufacturing practices and creates an environment where problems can be easily identified and corrected. Furthermore, the 5S system helps organize the work environment and keeps machinery and tools clean, organized and uncluttered; it helps to instill discipline in the workplace; and it sets the stage for more advanced techniques such as kanban and just-in-time manufacturing. Finally, it displays management commitment to an organized workplace.

It is the very first activity that is important in our quest for operational excellence. We tabulate in Table 2-1, the basic concepts of the 5S system.

First Step: *Seiri* or Sort and Organize

Here is an interesting exercise that may surprise you: Open your purse or wallet; how many items in there have never been used for months or should be discarded? Or open your desk drawer; how many items have not been used in the last 6 months? Check your filing cabinet: are there files, never used for 5 years that are still being kept?

Now, look at your production floor, how many machines, equipment, racks, pallets, or boxes are parked there but have not been used for a long time?

Therefore, the best way to start the sort process of sort and organize is to separate items that are required from those that are not used regularly. When identifying seldom used items we need to look at the following: Excess furniture, tools, equipment, operating supplies (such as chemicals), and WIP (work in process) inventory. Sort and organize will include the following:

1. For items that are used often but not daily, we need to ensure that there is proper place and method to keep them, so that they are easily retrievable. In this category are tools, light equipment that can be moved easily, and excess operating supplies and inventory.
2. We must minimize supplies and inventory on the factory floor. We can use simple inventory techniques to reduce these, but eventually more sophisticated techniques like Kanban will be required; more on this in later chapters. It is especially important to look for excess work in process (WIP) that is lying around, or waiting for repair or disposition.
3. For items that have little use or are used infrequently, we need to relocate them and ensure that they are retrievable in the future.
4. Finally, for items that are never used, we need to dispose, discard, or recycle them. Where appropriate sell or give them away.

Table 2-1: The 5S System

Item	Description
1. Seiri or Sort & Organize	*In this step we eliminate all clutter and unneeded items that have collected around work areas, warehouses, office space, and even the rear of factories.* As unused objects pile up, productivity is often impacted. The following steps are done and changes maintained: a. Identify unused items and move them to a designated area; for this activity we use the red-tag process. b. Excess items such as WIP (work in process) or WIP waiting rework must be minimized. c. There is proper location for all items, equipment, and tools in the factory. d. All workstations and areas are neat and organized.
2. Seiton or Stabilize or Set in Order	*Our goal here is to make it easy to find tools, equipment, and storage areas.* Specific activities in the factory must be identifiable and accessible. Hence we need to develop an environment that visually indicates where things and activities are: We must: a. Provide signage or coding to every area, equipment, entrances, rooms, production lines, and work areas to indicate the activity that is required or taking place there. b. Identify and mark floors and aisles to indicate the activity that is taking place: for example, ESD areas, safety warnings, walking routes, and production lines. c. Locate items in production lines and work areas to minimize motion and travel, and enhance productivity. d. Provide proper storage of inventory, material, supplies, and tools.
3. Seiso or Shine	*Once we have eliminated clutter in the work area, it is important to keep everything swept and clean.* Leaks, squeaks, and vibrations from equipment can often be easily detected, but a dirty workplace tends to be distracting and equipment faults can go unnoticed. Therefore: a. Ensure all areas are clean and shining through systematic and regular cleaning. b. Employees are actively involved in the cleaning process; checks are done routinely.
4. Seiketsu or Standardize	*The sort, set in order, and shine systems (discussed above) must be maintained via standards.* Therefore we must: a. Raise awareness of the 5S system and make employees aware of their responsibilities and expectations b. Decide on responsible parties to manage the 5S system in their own work areas. c. Integrate maintenance duties into regular work activities and job descriptions. Provide employees with standards and checklists to work with. d. Setup cleanup procedures and schedules for all areas in the factory.
5. Shitsuke or Sustain	*We must make a habit of maintaining all 5S procedures.* This is the most difficult "S" to implement and achieve. People tend to resist change, and even the most well-structured 5S plan will fail if not constantly reinforced. Begin by agreeing on the following: a. Get senior management involved in 5S system, meetings, and audits. b. Provide tools and resources to help measure progress and foster growth by conducting audits for 5S activities regularly. c. Find and implement ways to prevent the 5S system from deteriorating.

The Red-tag Process

To facilitate the sort process, we use the red-tag process. This is a method to identify un-needed items, review their usefulness, and to take appropriate action. We use a red-tag (because it's more visible) with the information shown in Fig. 2-1. On this particular red-tag we have provided basic guidelines to determine if and how much of an item should be red-tagged. The red-tag is used to record detailed information about the selected item. *One precaution: We will need a special label to indicate potentially hazardous materials.*

Besides removal of seldom used items, the red-tag process also requires that all items, tools, equipment, supplies, work in process, and raw inventory are properly placed on the floor; that is, there is a proper location for each item. Specifically:

1. There are no useless supplies, raw materials, semi-finished products, and waste in the work area.
2. There are no unnecessary tools lying around. Therefore all equipment, machines, and tools must be properly labeled, stored, and clearly marked.
3. Personal items, handbags, or lunch-boxes of production staff must be properly stored away from the production line.

Once we have implemented basic sorting and removed unwanted items from the factory floor and workstations, we can we can move to step two, which is stabilize or set in order.

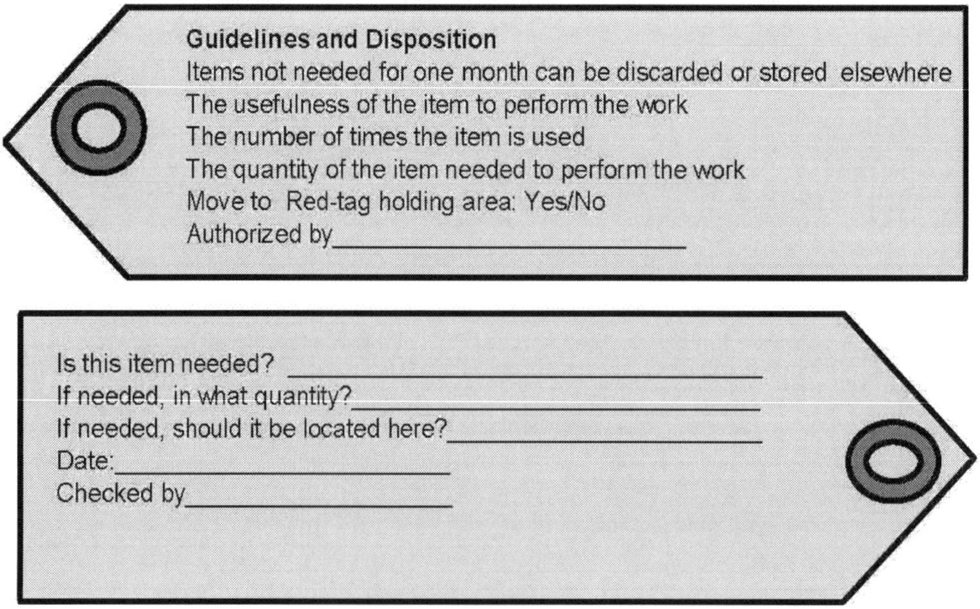

Guidelines and Disposition
Items not needed for one month can be discarded or stored elsewhere
The usefulness of the item to perform the work
The number of times the item is used
The quantity of the item needed to perform the work
Move to Red-tag holding area: Yes/No
Authorized by_____

Is this item needed? _____
If needed, in what quantity?_____
If needed, should it be located here?_____
Date:_____
Checked by_____

Figure 2-1: Red Tag

Second Step: *Seiton*, Stabilize, or Set in Order

Our goal here is to make it easy to find things. To ensure that, we need good signage and markings – hence the often used term: *The visual factory.* This is a factory that is clean, well organized, has little WIP (work in process), no visible rework within or at the end of the production line (or hidden in rooms), with good signage of all activity, and humming smoothly.

We want a visually attractive and organized factory. This means we must maximize utilization of factory floor space, with good layouts of equipment and processes, intelligent use of conveyor lines and storage racks. We will elaborate more on line layouts in later chapters. Some of the requirements for set-in-order are:

1. Re-arrange those items that we have decided to keep and store. However, we need to keep frequently-used items separately and make it easy to access and retrieve them.
2. Re-label and arrange drawers, shelves, cabinets, files, to allow easy access
3. Arrange workshop tools in easy to transport tool boxes. Commonly used tools can be arranged by shape and size on wall racks with shadow marking below tools for easy identification, access, and return.
4. Provide signage and labels to indicate specific equipment, pipes, cables, ovens, test and calibration rooms, laboratories, and specific production activity.
5. Initiate a labeling or marking system to identify and track your needed items. For example:
   a. By description
   b. By coding, e.g. alphabetical, monthly
   c. By shade, size or shape of objects
   d. By color coding
6. Minimize production material or WIP using a Kanban system.
7. Draw up or identify locations on floors and walkways to guide employees and highlight ESD (electro static discharge) areas and safety issues. Refer to Fig. 2-2 for sample coding guidelines for production areas. More comprehensive coding will be required for warehouses and loading bays.
8. Draw up floor markings for warehouses when they are insufficient boundaries or fencing.
9. In workspaces with many interior walls or large space it's best to use signs and labels to orientate employees and direct workflow. Examples include overhead maps giving current location and direction, and wall-signs.
10. All material on the factory floor must be easily identified and located for easy access.

Who's Minding the Kitchen?

Cleanliness and orderliness is necessary in every kitchen for very good reason. McDonalds is reputed for keeping its kitchen super-clean and organized, which is clearly visible from the order counter; certainly that helps to give the customer confidence in the quality, cleanliness, and hygiene of the kitchen and the food.

On the other hand, do you ever wonder what goes on in the kitchen of a fine restaurant? Actually, the French have got this worked out: To ensure that their kitchens are under control they have adopted the technique of *mise en place*: A French phrase which translates as "to put in place." Mise en place requires scrupulous assembly of all of the ingredients, pots and pans, plates, and serving pieces needed for a well-managed kitchen.

According to the Culinary Institute of America's manual, '*The New Professional Chef*': "Someone who has truly grasped this concept is able to keep many tasks in mind simultaneously, weighing and assigning each its proper value and priority."

(Source: Wikipedia).

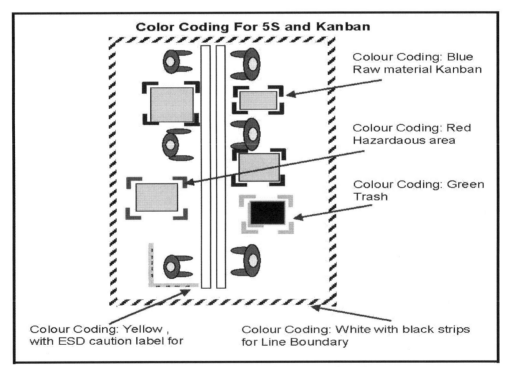

Figure 2-2: Sample 5S floor coding for a production area
Here we illustrate sample line coding for a production area. This identifies locations on floors and aisleways to guide employees. Areas are marked out for various trash, ESD zones, kaban deliveries, and boundries for production areas.

Third Step: *Seiso* or Shine

After we have eliminated the clutter in the work area and improved the layout, it is necessary to thoroughly clean the area and the equipment. Despite all the hard work of keeping things clean, they will get dirty again due to dust, dirt, humidity, temperature, usage, and misuse. Hence this third stage of 5S is important to ensure there is a regular cleaning process of all areas and equipment in the factory. It is crucial to remember that all employees will have to be actively involved in the cleaning process; for many of them, this becomes part of their daily activities.

Therefore we start by ensuring everything is clean and shining, through systematic and regular cleaning. The cleanup and shine process will help to uncover many issues and problems like:

1. Leaking water pipes, growth of rust, defective machines, corrosion, spilled oil and liquids, or decaying floor coverings.
2. Poor processes such as equipment that is seldom cleaned, service personnel that leave mess behind, packaging material that create dirt and debris, and transport equipment that causes damage and creates dirt.
3. Areas that are always dirty and continue to attract dirt. Typically, problems on clean equipment can be easily detected during routine cleanups, but a dirty environment will distract attention and equipment faults may go unnoticed. In fact, during the initial cleaning activity, employees will notice areas that seem to attract garbage or spills. It is important to identify such areas, and pay special emphasis to ensure deep cleaning. Such areas may also indicate machine and equipment faults, leakage, or corrosion; hence technicians will need to be called to fix the problems to prevent recurrence.

For every item it is important to determine the method, frequency, and ownership for cleaning. It is essential that proper tools for cleanup are provided. Hence we must ensure that:

1. Cleaning equipment is readily available.
2. Cleaning supplies are replenished routinely and marked with instructions.
3. Areas containing cleaning equipment are properly marked.
4. There are sufficient waste receptacles, which are clearly labeled for different types of waste: Disposable waste, production waste, and chemical/toxic waste that will require special disposition.

Finally, remember that a clean workplace is important for employee health, morale, and safety. These factors add up to better productivity and improve the company's bottom line.

16 *Winning with Operational Excellence*

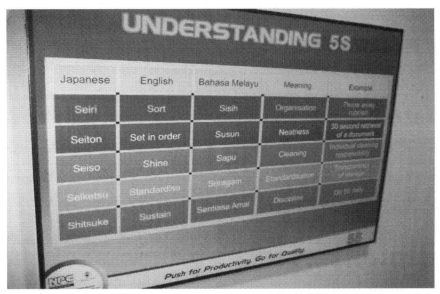

Figure 2-3: A 5S Display Board
Here the company is publicizing and explaining the 5S system. This company also conducts 5S audits and has a reward and recognition system. Figs. 2-4 and 2-5 show good 5S activities at this company.

Fourth Step: *Seiketsu* or Standardize

Unfortunately, cleanliness does not last forever. Consequently cleaning and organization systems must be implemented with standards and goals to sustain the momentum of 5S. Hence, we must:

1. Raise employee awareness by conducting 5S skills-training and setting expectations.
2. Decide on responsible parties for each area or equipment. Posted work instructions or standards must list cleaning activity required as part of the overall work procedure at each workstation.
3. Have an overall cleanup checklist that looks for issues that we have addressed in the first three steps of 5S: organize, set-in-order, and shine. Specifically:
   - Always look for old and unused tools and equipment on the production floor. These may have accumulated since the initial red-tag clearance.
   - Ensure signage (sign boards, floor markings, labels) is well maintained and all shine and cleanup procedures are implemented
4. Integrate maintenance and cleanup duties into regular work activities:
   - Establish rules and regulations.
   - Educate employees on awareness of 5S.

- Display Do's and Don'ts.
- For each work area we must provide short checklists for daily use.
- Cleaning standards should be clearly displayed around the workplace using signs and posters or in the work standards. Note that established and published standards offer employees and employers a way to reach common goals while showing impartiality to both sides.

Fifth Step: *Shitsuke* or Sustain

The fifth step of the 5S system is to sustain everything that has been done so far. Sustaining and maintaining the 5S system is the most difficult step; Why? People tend to resist change, and even the most well-structured 5S plan will fail if it is not constantly reinforced. This step to sustain is better termed as *discipline*. So we need to set up specific objectives to maintain the 5S momentum and ensure that we do not revert back to bad old habits. Specifically:

1. Set goals for the 5S system to maintain the current status, and measure ongoing status.
2. It is crucial that we provide tools and resources to help measure progress and foster growth. For this step, we recommend doing a 5S audit.
3. Ensure that improvements that are identified during a 5S audit are implemented.
4. Publish results to motivate the team, take corrective actions, and establish a reward system for good work done.
5. Reward systems can vary from monthly recognition events to yearly cash or gift rewards for best team effort; we have found that monthly recognition events are well received by employees. In all cases, the recognition process has to be transparent and communicated to all employees. The 5S audit and scoring system discussed later.
6. Select areas to improve: This can be departments that are weak – based on impartial audits – or specific areas that require attention such as the rear of a factory. This step if done correctly will identify and reinforce the discipline required to sustain and improve the entire 5S system.

Fig. 2-3 to 2-5 shows some examples of good 5S practices at a power station[10]. Power stations can be poorly maintained and dirty, but this one is so clean and shining, and if you wish you can have a meal sitting on the floor. Fig. 2-3 shows a 5S display board to publicize the 5S system, while Fig. 2-4 and 2-5 show very clean and organized areas. It is very clear that both management and operators at this station have been able to implement an excellent 5S system. Fig. 2-6 shows a clean and well-laid out production line in a factory.

Figure 2-4: Shining and organized areas in a Power Plant

Figure 2-5: Organized tool-storage area and color coded garbage cans

Sustaining 5S in the Factory

What are the essential ingredients for successful 5S? Once the 5S system is started, it will need to be promoted, encouraged, and cultivated until it is well established. We list several ingredients that have worked:

1. Top Management support: The Company's top management must set the expectations and benefits of the 5S effort. To this end management must set up a 5S committee to help facilitate and drive the process. Specifically:

   a. This committee must be setup at the very beginning with participation of senior management, to indicate and ensure total management commitment. Ideally, the factory or operations manager should be the committee chairman. With time and success, the activity can be managed by lower level managers and employees. The committee should include representatives from all departments in the factory.

    b. This committee will ensure publicity, communicate expectations of the 5S effort, and set up a system of rewards and recognition for good work. Ideas for the reward and recognition system were discussed earlier in *Shitsuke*.

2. All employees must be encouraged to participate. From the manager in keeping his or her office organized, to the technicians who keep their equipment and area clean and organized, to the shop-floor operator as part of her daily job. Employees like to be involved in company activities, especially in an improvement program, and the 5S effort is no exception.

3. New employees must receive 5S training in their orientation session after they join the company and start work. Yearly re-training of employees should include improvements and new challenges for the ongoing 5S effort.

4. The committee should appoint a 5S audit team to implement 5S audits, which will include committee members. The team will start a routine 5S audit with checklists and scoring.

Figure 2-6: A clean and well-laid out factory floor.

Conducting 5S audits

To help sustain the 5S effort, we need to establish a 5S committee to help facilitate the entire 5S process and drive improvements. Basic 5S theory will get us started. However, further improvements can be identified by using a good and structured 5S audit process. An audit team should be identified to conduct the audits, and should include 5S committee members and employees from each operation. The team will start a routine 5S audit with checklists. To get started we recommend:

1. Setup a 5S audit team. The team should comprise of employees from all relevant departments. Note: This process should not be left to the quality or HR department.
2. To get started, it is best to get a senior manager to steer and drive the audit team; this will show management commitment.
3. Team members must have been trained to understand 5S and given time to conduct the audits. We recommend weekly audits.
4. It's best to conduct the audits randomly and at different times each week.
5. In addition to the weekly audit, employees in each area should be required to do short 5S audits, including daily clean-up, at the end of each production shift. This will leave a clean and well organized line for the next shift.
6. In Table 2-2, we show an audit checklist for a factory floor. The list breaks down expectations for each of the 5S categories into detail. Typically such a report must be accompanied by photos of the good and bad areas observed during an audit, plus list the responsible managers or supervisor to fix each issue.

In Table 2-3 we show an extract of a 5S audit photo report. The photos and comments will be used to finalize the 5S audit report in Table 2-2. It will be obvious from the report that the auditors are going beyond housekeeping: They are identifying underlying problems, such as:

- Too much inventory: Evidently guidelines for kanban and inventory control are not adhered to.
- Potential ESD (Electro Static Discharge) failures because material storage guidelines are ignored.
- Dangerously placed parts, which can lead to damage and quality issues.
- A dirty and disorganized work environment that can impact quality of repair operations.

With regard to the last audit comment above on "dirty and disorganized work environment", employees should be trained to keep their workstation clean and organized. To facilitate employee audits of their workstations, we recommend a short checklist for the production operators which they can use in each work area; such a list can be a shortened audit list, derived from the formal weekly 5S audit list.

Hence we reiterate that one of the objectives of 5S is to create an environment where problems can be easily identified and corrected. Furthermore, when 5S is executed efficiently, it sets the stage for more advanced activities such as kanban and just-in-time manufacturing.

Table 2-2 5S Audit checklist. Reviewers:		Date:	
Item	**Scoring Criteria:** Rate each category item with score of 0, 0.5, or 1. Maximum score per category is 5. Overall per category rating will be score of 1-5 (poor = 1, excellent = 5).		Score
1. Seiri or Sort	a. Workstations should be organized and neat, with only the tools & products necessary to perform tasks. b. All unneeded items are red-tagged and isolated pending disposition; note that red-tag process is an ongoing activity. Seldom used, but needed items need to be identified, stored, and accessible. c. There is no useless or excess WIP (work in process), products, or waste in the work area. d. There are no unnecessary tools lying around. All equipment, machines, and tools are properly stored and well organized. e. Are personal items of workers properly stored and not on the line?		
2. Seiton or Set in Order & Stabilize	a. All jigs, fixtures, tools, equipment, & inventory are properly identified and in their correct locations. Ensure correct locations have been identified and marked. b. Only documents necessary to do the work are stored at work stations and are in a neat & orderly manner. c. Floors, aisle-ways, and workstations are clearly marked. Metal guards & deflectors are in place at critical or weak points along aisle ways. d. Are lines, labels and signs updated and in good condition? There are clear markings as to where WIP and other items are stored. e. Notice boards are displayed neatly and the information is *updated*.		
3. Seiso or Sweep & Shine	a. All areas are clean and spotless; Equipment and machines are clean – front & back. Nothing is placed on top of machines & cabinets and their tops are dirt & dust free. b. Areas prone to dirt, leaks, spills, corrosion are clean and well-managed. c. There are no pallets, boxes, and loose items on the floor preventing cleanup. d. Cleaning equipment is available and stored in a neat manner and easily accessible. e. There is a good process to ensure that no dust and dirt is brought in from outside into the production area.		
4. Seiketsu or Standardize	a. All employees have been educated and aware of the 5S system, goals, and progress is communicated routinely. b. Employees in each section (production, maintenance, etc.) have a dependable, documented method of preventive cleanup. c. Cleanup procedures are part of each section's daily work procedure, and they perform routine daily cleanup. d. Cleanup procedures, standards, or checklists are posted at all critical workstations. e. The 5S standards are entrenched in the work environment; and the operation is continually seeking better ways of preventing 5S problems.		
5. Shitsuke or Sustain	a. The operation is absolutely spotless and all machines, equipment, work areas, glass surfaces, and buildings are painted or kept clean by routine daily care. b. A high degree of organization is evident, everything is in its proper place, no excess material or WIP; material on the floor is minimized and managed via Kanban. c. Goals and progress of 5S is visible and progress are evident. d. Everyone knows what the 5S's are; there are routine 5S campaigns each month to promote progress. e. This is clearly a World-Class operation.		
	Total Score =		
	Normalized Score in % = Total Score x 4		

Table 2-3: Line 1: 5S Report & photos

Area Reviewed:		Reviewer:	Date:	
No	Photo	Description	Counter-measures	Who
1		Station L1-07: Case assembly. Excessive raw material (top case) is over the Kanban Square limit. Material is spilling over the Kanban lines. Some material is too close to the conveyor, exceeding the ESD guidelines of 30 cm. minimum distance.		
2		Station L-23: Assembly station. We observed motor parts improperly placed in their containers; some are in danger of falling to the floor.		
3		Station L1-36, Debug Station: Work area is disorganized and untidy. Area and equipment is not clean and can transfer dirt and debris to repair assemblies or products.		

Tracking 5S Progress

How do we measure progress in the 5S system? The 5S audit procedure in Table 2-2 includes a scoring method. The scoring system is not perfect, but is able to give a ballpark score of current status and show progress over time. After each weekly audit a score can be computed and progress tracked. Typically there should be a 5S goal, of say 80%, which every production line strives to achieve. Fig. 2-7 is an extract of an operation's 5S score.

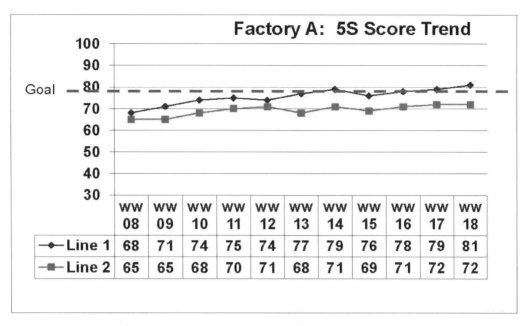

Figure 2-7: Tracking 5S Progress via a Scoring System and Goal

Barriers to a Good 5S System

In spite of hard work to create a good and effective 5S system on the factory floor, there will be bad days due to lack of enthusiasm and just bad days. What are some barriers?

1. *Awareness and training of the 5S system are lacking*: Therefore we need to reinforce training and education.
2. *Sometimes, employees simply ignore rules*: This implies that staff and operators need more education to understand their roles and supervision to comply with procedures.
3. *No time to meet or implement 5S,* because meeting shipments is the number one priority: This is a difficult issue, in such cases it's best to work on 5S during the first half of a month.
4. *The 5S audits are not conducted or not comprehensive enough to surface issues;* worse still no corrective action taken after an audit. This is a management issue and they need to meet and resolve.
5. *The 5S system was implemented years ago but change of management and lack of motivation has caused the system to fragment*; in fact we can still see faded 5S posters on the wall. This is a common observation and clearly a management issue. Per Dr. Deming one of the Seven Deadly Diseases of Management is "lack of constancy of purpose". Therefore, management needs to reinforce the goal and practice of 5S: Employees follow what their leaders emphasize or do.

We can illustrate potential barriers and solutions by looking at a common problem: Unused or broken equipment on the production floor or on shelves. This is shown in Table 2-4. If the potential solutions are reviewed and implemented, many of the barriers to success will be removed.

Table 2-4: Barriers and solutions for unused or broken equipment on the production floor	
5S discrepancy observed	**Potential solution**
1. No proper storage for broken items, hence no one knows where to take broken stuff	Select and identify a space to store unused and broken items. This is classified as the red-tag disposition area, and should be a permanent location in every factory.
2. No clear rules and procedures for clearing broken stuff.	There should be a routine red-tag exercise at the factory; it can be done monthly as part of the formal weekly 5S audit. All work-sections should have their own weekly cleanup process that also looks at unused and broken equipment.
3. New employees accept poor housekeeping as a norm and are not aware that it's unacceptable.	New employees should go through 5S awareness training. This should be a part of the new employee training program, and will foster and create employee involvement and success.
4. There is no consequence for inaction or employees are lazy and not motivated to clear broken stuff.	Supervisors must communicate department goals to employee and set expectations; if there is a repeat of such incidents, disciplinary action is required.
5. Management tolerates such activity and accepts a poorly managed production floor – their focus is "Shipment first"	This is the most difficult item to fix. The management team, and specifically the operations head, needs to resolve this and determine if this is a priority and part of their quest for operational excellence.

Dear

Thank you for hosting us on Thursday, September 6th 2xxx. Out group appreciated your team's time showing us the capabilities of your organization.

We were impressed that you have a good understanding of NPI's unique requirements; your team was more than willing to address areas of concern and showed a willingness to modify or change processes to meet NPI challenges. However, there were some areas of concern that must be improved & enhanced to meet our standards and to meet current Industry Standards for IPC-A-610D Class 3: Specifically:

ESD Program

    Drag Chains (stock-room work station drag chains sitting in pool of dry spilled liquid)
    Tables had grounded wires dangling not hooked up
    ALL racks and shelves were not conducted properly
    Employees were not grounded when handling boards/components

Component Control

    Excess and staged kits are kept together
    Floor/machine personnel are responsible for labeling their kit parts
    Loose parts in a bag were found on the kit/excess shelf unmarked

Visual Aides

    SMT hand placements on each line – no visual aide or bin marking
    T/H hand stuff – no visual aide

Documentation Storage

Process ID dependent on production floor personnel checking to ensure that they have the correct revisions documents. This is normal but typically document control drives documentation release and retrieval.

The Manufacturing Engineer seemed disconnected to some of the processes

    Didn't know why hand placements were being done
    Didn't know the DI water resistivity level for his cleaner

Equipment

    Machine areas were dirty and not well kept
    Work stations were old and not well kept
    Current in-process documents and procedures were out of date
    Calibration stickers were out of date on equipment in use
    Room temperatures and humidity monitoring not in use although a device that monitors temperature and humidity was in place

In summary, we were encouraged during the initial kick-off meeting. You may have the right answers to meet our needs. However, during the walk through we found the above noted areas of concern. We recommend you obtain outside assessment and assistance regarding any requirements for Class 3 requirements. If you remedy these concerns as validated by independent review, then you should contact us to be reconsidered as a potential Supplier.

    Sincerely, Procurement Manager

Figure 2-8: Customer feedback letter after a factory tour

Do Customers Really Expect to See an Organized Factory?

The answer should be a resounding yes. Still, we were astonished to see a letter, which a sales manager shared with us. We show in Fig. 2-8, an extract of a letter from a customer team, who visited a factory that was bidding for new business. Sensitive information has been deleted from the letter, but all the comments are unedited; we can't make this up!

Beyond board room presentations (which are often very impressive), the purpose of a customer tour is to judge the factory's ability to build products efficiently and with high quality. Apparently the customer was not happy during the visit, but took the trouble to document his detailed observations, as shown in his letter to the factory manager. The customer's letter reflects his observations over a short visit and mentions several major issues:

- Poor housekeeping
- Unacceptable condition of tools and equipment; in addition un-calibrated machinery: This is an indicator of sub-standard equipment and inefficient processes
- Excess work in process inventory.
- Lack of ESD awareness and standards.
- Staff not aware of basic manufacturing procedures.
- Inadequate training of staff; this is reflected in the lack of visual aids, training material, or work standards.

The issues mentioned in this letter, such as basic cleanliness, organization, training, and standard work are addressed in this and succeeding chapters.

Summary: The Basics – Getting Organized

We have discussed the 5S system, the detailed steps of 5S, how to sustain 5S, conducting 5S audits, and how to overcome the barriers to 5S.

The 5S system lays the foundation for good manufacturing practices and creates an environment where problems can be easily identified and corrected. Furthermore, the 5S system helps organize the work environment and keeps machinery and tools clean, organized, and uncluttered; it helps to instill discipline in the workplace; and it helps to set the stage for more advanced techniques such as standard work, kanban, and just-in-time manufacturing.

From senior management's perspective the 5S system will provide a strong foundation in the factory environment and is the first activity that will test organizational readiness for operational excellence and lean manufacturing.

When 5S is executed efficiently, it sets the stage for more advanced activities, such as the high performance factory; that's our next topic.

Chapter 3

High Performance: Standard Work

Three Rules of Work:
Out of clutter find simplicity
From discord find harmony
In the middle of difficulty lies opportunity.
Albert Einstein

Where there is no standard, there can be no kaizen.
Taiichi Ohno

Overview

Every manufacturing operation must have good work procedures for the products it manufactures. The procedures have to be well documented and training provided to its employees to ensure good and repeatable processes in manufacturing. Such detailed and comprehensive work procedures are termed standardized work or standard work. Good, accurate, and well-prepared standard work will lead to predictable high performance from production operators.

In this chapter we will discuss standard work, specifically:
- The content of standard work.
- How to prepare it.
- How to use and improve it.
- Review the importance and methodology of good operator training.

Preparing and Implementing Standard Work

Standard work documents the current methodology for building a product in the most efficient way with the best quality. If the standard work is implemented with operators trained correctly, then the work will be done in the same best way – no matter where the work is done: another work-shift, factory, or geography. The emphasis here is on the best and most efficient way of doing work on a particular job. After the standard work is determined, operators are not allowed to deviate from the standard, except when they have the opportunity to discuss and improve it.

Written work instructions have been essential requirements since Frederick Taylor wrote his *Principles of Scientific Management*[11]; since then, however, they have improved in format and content, to include operator involvement. Henry Ford[12], when he ran Ford Motor Company, also believed in the benefits of standardization and said that it laid the foundation of tomorrow's improvements. Toyota Motor Company has the same belief and insistence on the importance of standard work. Hence, the industry leaders are very clear that documented and implemented standard work will maintain processes at a stable level, but more important they will form the foundation for driving continuous improvements on the factory floor. Standard work is a requirement in every manufacturing environment.

In addition, every manufacturing operation must have a comprehensive and effective skills-training program. The purpose of such a program is basically to prepare all production operators to be skilled, productive, and efficient in all the work they do. Besides generic training and awareness of company policies and general skills, they must be trained in specific skills, or standard work, to build products on the factory floor.

Content of a Standard Work

Standard work defines the methodology for work to complete one job, work, or process in a multi-process environment. Essentially this means that any production operator doing the same work, during any work shift or at any factory, will use the same method. The current work begins after the previous work sequence is completed and ends before the next work sequence begins.

What must we include in standard work? There are three components:

1. All the steps in the work sequence: This is discussed next, in detail.
2. The workstation cycle time to do the work: We will cover this here, but a detailed discussion on managing cycle times in production will be done in the next chapter on *managing cycle time*.
3. The material and inventory required on hand for the work: We will cover this here, but a detailed discussion will be done in the chapter on the *kanban system*.

Let's look at a comprehensive list of requirements for standard work. This will include the sequence of work, the material that will be used, and the tools used to perform the work. The specific requirements for standard work are:

1. The process steps:
    a. **The document title.**
    b. **The work instruction in correct sequence** at each station must be clearly stated. Where possible the work instruction will use photos to illustrate the work required. A picture is worth a thousand words – hence the photos or cartoons must be very explicit and helpful to the production operator.

      c. **Tool(s) required** for performing the job, such as screwdrivers, power drills, wrenches, press, or measurement gauge.

      d. **Personal requirements** of the production operator, such as gloves, work-smock, safety shoes.

      e. **Quality information** that is specific to the job. This can refer to a specific instruction that will prevent a defect. For example: how to lay a wire-harness bundle in order to avoid potential damage or describe how specific work should be inspected.

      f. **Safety information** can refer to a method or warning to the operator: for example to wear gloves to avoid an electric shock and to ensure gloves are in good condition.

      g. **Observation log:** For complex jobs, it is advisable to provide a log book for the production operator to record difficulties encountered and suggestions for improvement. This ensures feedback to the production supervisor and engineer on how to improve the process and revise the standard work.

      h. **Other technical details,** like the author, revision date, and approval.

2. **The cycle time:** The time it takes to perform the job. Depending on the specific job performed, the cycle time will include the actual time to perform the job and non-productive time such as time to taken to move or walk at the workstation.

3. **Material and inventory required for the job:** This is the list of parts that are added to the work during the job. Hence, we need to list the parts and quantity required during assembly. Where possible, photos of each part should be included; this ensures the production operator can identify the part visually and not just by part number.

4. **Recommended inventory level** of parts at the work station: This refers to the inventory of parts required to be stocked at the workstation which will subsequently be added to a work-piece as it arrives at the workstation or the number of assemblies if the work is performed on a batch of assemblies.

Sample Standard Work

In Fig. 3-1(A, B, and C) we show an extract of a standard work for a mechanical assembly. The work instructions are documented in detail and are used to train the operator on how to do the assembly work sequence. In addition, cycle time information is provided in Fig. 3-1C. Note the following points:

    Written work procedures: There are written work instructions and pictorials for each step. In addition, there is a record log provided at step 1, where problems and suggestions can be entered for later review by the production supervisor and engineer. This is very low tech, but it is a simple and effective tool for continuous improvement. The standard includes a list of parts required for the job, a list of tools required, tool settings, and specific safety and personal dress-code requirements for the production operator.

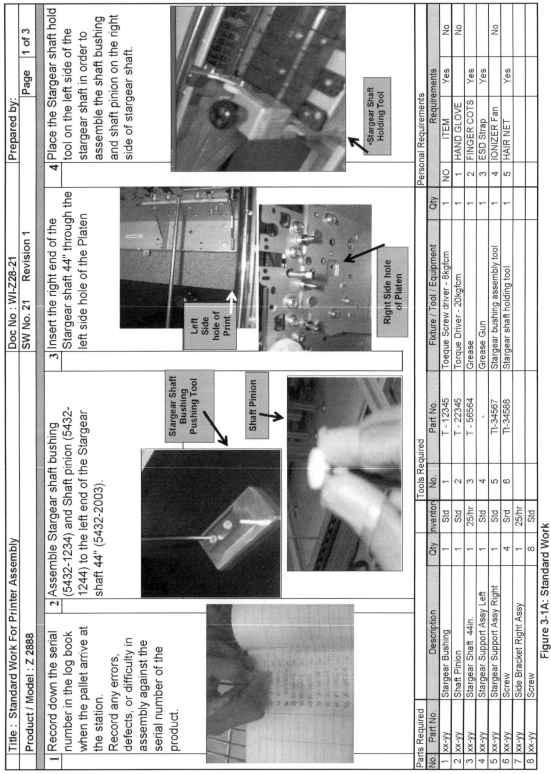

Figure 3-1A: Standard Work

Inventory of parts required for the station are listed: In this particular workstation, there is a requirement for a standard inventory lot for low cost and non-bulky items, which typically is a full bin of parts. This is replenished a few times a day by a material handler. The other requirement is for expensive or bulky parts: 25 units per hour in this case. This means the kanban process, will deliver 25 units to the station every hour. As the cycle time is 175 seconds, this means slightly more than an hour's production requirements are delivered every hour. The person who delivers the kanban requirement is tasked to ensure no excess parts build-up occurs at the workstation. Often the kanban guidelines are managed separately by the parts delivery team; this is because of the need to update the kanban process when production volume varies.

The station cycle time is computed and listed in Fig. 3-1C, and includes total work time T, comprising of M (manual or queue time), P (process or work time), and W (walking or wait time).

Preparing Standard Work

Preparing the standard work requires an experienced engineer or technician with a trained eye to observe and improve the process. Creating an effective standard work is an ongoing process requiring continuous improvement, and is not a one-time project. Hence we start with an initial draft. We will not be able to tell if the standard work is effective until we implement it.

Typically, a draft is prepared together with the process engineer and the R&D engineer involved in the initial design. This is done prior to new product introduction (NPI). The assembly steps are projected per the work step(s), the bill of materials required, and the planned cycle time. The required cycle time is dependent on the production volume planned for the product – this is discussed later in the chapter on cycle time.

Preparation of the work instructions and the cycle time at each station should proceed as follows.

1. The engineer needs to record the detailed assembly steps exactly and in the sequence that the work must be done. The sequence is very important as in certain designs the wrong sequences of operations may cause errors.
2. Next the cycle time for each station must be measured. The best way to do this is to break up the work procedure into discrete individual steps. For example, the typical steps at a workstation could be as follows:
    a. The operator reaches for the work piece or job.
    b. The operator prepares to work on the job, reaches for tools, gets parts.
    c. The operator works on the job.
    d. The operator loads the job into a machine or tester.
    e. The operator completes the job and puts it away.

32 *Winning with Operational Excellence*

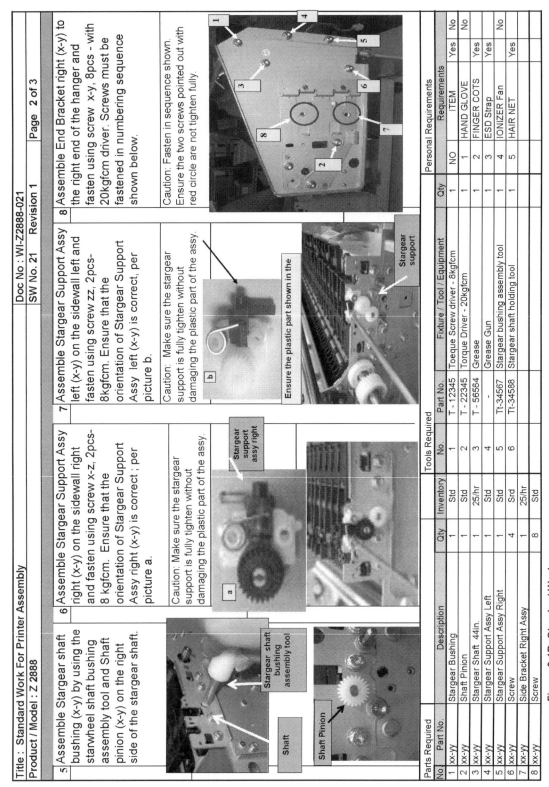

Figure 3-1B: Standard Work

3. **Definition of cycle time**: Let's take a break and define cycle time:

   Cycle Time = arrival time + queue time + setup time + process time + batch time. Where:
   - **Arrival time** is the time to move a job from the previous station and includes walking time
   - **Queue time** is the time the job is waiting to be processed or to move to the next station (after processing)
   - **Setup time** is the time it takes to set up the equipment for doing the job
   - **Process time** is the actual time to process the job at the current workstation
   - **Batch time** is the average waiting time to batch the job into a quantity of more than one unit. This would be zero for one-piece operation.

   Looking at the cycle time definition, it is very clear that the process time is the only productive time, while everything else is waste or *muda*. Hence, in preparing standard work, the role of the engineer is to minimize muda. Fig. 3-1 C, shows cycle time T and its components: Where, M = Manual or queue time, P = Process or Work time, and W = Walk or Wait time.

4. It is important to collect time data accurately and it's best to measure each discrete element of the work sequence.

   a. An element can be considered as the tiniest visible work in the process. If the elements are captured accurately, they will help to identify future opportunities for improvement.

   b. Measure the time data using a stopwatch for each element and record onto a worksheet. Plan to record at least 10 full cycles of assembly at each station by elements; then use the average as the target cycle time. If the variations around the average are large, then check why this is so. When some recorded cycle measurements are much higher than the average, the variation may be due to material shortage, test issues, or quality issues -- check for them.

   c. For a complicated work sequence, it is appropriate to make a record with a camera for later review.

   d. An example of recorded cycle time is shown in Fig. 3-1C. This is a work combination sheet with all the work sequences measured and recorded for the work station in Fig. 3-1A & B. We should record each element separately and not merge into chunks of work. For example, step 1 in Fig. 3-1C shows two elements: manual and walking. Often for simpler assembly procedures, only the overall time is recorded.

   e. Note the actual process time Vs all other components, which are *muda*. This *muda* component must be reviewed and reduced if possible.

34   *Winning with Operational Excellence*

f. While Fig. 3-1C displays the work done at one workstation, Fig. 3-2 displays work done at a cell with five workstations. Here the operator goes through five steps to complete the job.

Title: Standard Work For Printer Assembly					Work Combination Sheet					
Work Station or Cell: SW No. 21		Product: Z 28			Line: 12		Prepared by:		Doc No: WI-Z28	
TAKT Time: 185 secs		Cycle Time: 175 secs					Date:		Page 3 of 3	
M=Manual/queue; P=Process/Work; W=Walk/Wait; T=Total					Operation Time (seconds)					

Step	Step / Element	M	P	W	T	Operation Time (seconds) 10-180
1	Pull pallet, position, record Serial No in log book	8		8	16	
2	Assemble Starwheel shaft bushing and Shaft pinion to the left end of the Stargear shaft 44"	6	17	4	27	
3	Insert the right end of the Starwgear shaft 44" through the left side hole of the Print Platen	4	9		13	
4	Place the Starwheel shaft holding tool on the left of the starw shaft; assemble the shaft bushing and shaft pinion on the right side of the shaft	6	11		17	
5	Assemble Stargear shaft bushing by using the star shaft bushing assembly tool and Shaft pinion on the right side of the starwgear shaft.	5	18		23	
6	Assemble Starwheel Support Assy right on the sidewall right and fasten using screw, 2pcs- 8 kgfcm. Ensure that the orientation of Starwheel Support Assy right is correct .	8	14		22	
7	Assemble Starwheel Support Assy left on the sidewall left and fasten using screw 0515-4793, 2pcs-8kgfcm. Ensure that the orientation of Starwheel Support Assy left is correct, per pix a.	5	11		16	
8	Assemble End Bracket right to the right end of the hanger and fasten using screw 0515-4862, 8pcs - with 20kgfcm driver. Screws must be fastened in numbering sequence shown in daigram	6	22		28	
9	Push out pallet, turn, and push to right on conveyor	6		7	13	
	TOTAL	54	102	19	175	

Figure 3-1 C: Work Combination Sheet.
The figure shows the detailed station cycle time for the work sequence in Figure 3-1A & B

Title: Standard Work For Print Chassis					Work Combination Sheet						
Work Station or Cell: 2					Product: Pam		Line: 1				
TAKT Time: 70 secs					Cycle Time: 66 secs						
M = Manual or Handling; A= Machine or Work; W=Walk; T=Total					Operation Time (seconds)						
Step	Step/Element	M	A	W	T	10	20	30	40	50	60	
1	Remove sheet metal from incoming & load into press and cut	6	4	2	12	■						
2	Move workpiece into press and bend	5	4	3	12		■					
3	Drill 16 holes	4	9	4	17			■				
4	Insert 4 connectors and press fit	10		4	14				■			
5	Clean and Polish finished workpiece and place in rack	7	0	4	11					■		
	TOTAL	32	17	17	66							

Figure 3-2: Work Combination Sheet for a work cell with five separate workstations

5. Before the production run, it is important that the production operator is selected for each work station based on job complexity and operator experience and education.

6. Production is then run and the actual performance is observed and the draft standards are observed and validated during the assembly process for the entire production line.

7. Note: In the appendices, we provide additional information and access to a worksheet for standard work preparation.

Improving Standard Work

The data collected during the prototype stage will be used to fine tune the standards for the entire production line. Hence, we need to observe each and every work step in detail. Before the improvement process starts, we should allow some time for the operators to move up the learning curve.

1. **Initially only gross discrepancies are fixed, one by one.** Once production starts, the standard must be reviewed and improved.

2. **Determine if the actual work performed is different from the original draft** standard and understand why it is different; talk to the production operator to discover the reasons for the deviation and any difficulties incurred during work. This step is very important as the standard work must involve inputs and suggestions from the production operator who did the job at each work station.

3. **Observing and analyzing work at each station will allow us to discover activities that are not adding value** and hence must be eliminated or improved. Such activities can range from a production operator having to leave a work area to obtain parts or tools, to doing excessive physical movement, to difficulties in operating machines or equipment.

4. **Reducing motion and travel with spaghetti chart analysis**: One of the potential areas for improvement is unnecessary, transportation, travel, or motion. If an operator works beside a conveyor line, her position may require little travel or motion; even so production parts, material, and tools should be easily accessible so as to minimize travel and motion. But if she works in a cell and is required to travel and access equipment, parts, and several processes, then travel and motion may be creating waste or muda. In such a case, an engineer needs to review the station or cell layout and reduce travel time.

    This is done by drawing a spaghetti chart over a workstation, or cell, layout – this will display the travel routes taken by the operator. On the chart, excessive travel will be clearly visible. Here are some guidelines for drawing a spaghetti chart:

    a. Review the current standard work and draw the current layout of the workstation or cell.
    b. As the operator works on her job, chart the timing of each work activity and the travel routes. Sketch the travel route with spaghetti lines.
    c. Calculate the distance travelled, time taken, of each activity. Discuss with the operator to understand the reasons for the travel.
    d. Create a separate copy of the layout and write potential improvements on the chart.
    e. Consider all the following improvements: Reorganize the workstation (proper location of parts, golden samples, tools, waste bins, work-standard); review travel routes and re-layout of jigs and equipment to shorten travel distance; install smaller bins and racks which are less cluttered and easier to access – this smaller quantity of parts must be supported by a kanban process.
    f. Review the proposal with the operator and supervisor and plan the improvements that are practical.
    g. Fig. 3-3 illustrates such an improvement procedure. The figure shows the before and after layout of the workstation. The distance and timing improvements are summarized in the table
    h. Note: It's best to minimize the re-layout effort by designing the initial standard work with compact work-stations and to ensure that the work-piece is within easy reach of production parts, tools, and equipment with a minimum of operator travel.

5. Also review the work done and cycle times at each station and decide how to equalize or balance the cycle times at all stations – a more detailed discussion is provided in the chapter on *cycle time*.

Improving standard work via good observation requires patience and a trained eye. It requires observing the work at hand and understanding the current difficulties and ways to improve. This technique of observing is known as *genchi genbutsu,* which is Japanese for "go and see"[13] and is an integral part of good factory management. It refers to the fact that any information about a process will be simplified and therefore lack accuracy from the original when reported through another person. This is one of the major reasons why solutions designed with insufficient understanding of the actual process may end up being inappropriate or wrong. Therefore, to understand the real problem, you must go see for yourself – at the location where the activity is occurring. Once you understand the detailed process and its challenges, you can implement a good and effective improvement.

Genchi Genbutsu: The Gemba Walk

Genchi Genbutsu is an important component of the TPS and is about making decisions after observing and getting your hands dirty by spending time on the production floor. Managers, supervisors, and engineers need to take this approach when they work on improvements of standard work, or on process problems highlighted by customer feedback.

This approach of genchi genbutsu is also called *gemba*, which is Japanese for "the place", or where it's happening. The gemba approach – often called *the gemba walk* – is about understanding and making decisions after observing and getting your hands dirty on the production floor.

Managers and engineers should make it part of their daily routine to observe activities on the shop floor. A gemba walk can help them discover waste and identify potential areas for improvement. Here are the steps to do an effective and productive gemba walk:

1. Go observe the actual situation and understand the processes you are observing.
2. Ask questions to clarify your understanding.
3. Look for documented and posted standards, with visual work instructions: Understand the work flow, cycle time, and inventory on hand. Look for deviations, disruptions, and wasted effort.
4. Check to understand that the standards are clear. Do the operators understand them? Often, standards are done to fulfill management objectives but no thorough training or thought may have gone into them.
5. Check when the standard was last changed. An important question to ask: "Have there been changes and updates in work standards for this production line?" If no changes for a long period, ask the question: "Is this line running with perfect work standards?" Answers to these questions will reveal many issues and provide deeper understanding.

6. Collect and document data, take photos to verify where possible.
7. Look for excessive travel, motion, or difficult procedures and discuss with the operator. Document your findings.
8. Evaluate and analyze all the data and facts, understand what works or does not work.
9. Decide on specific improvements and next steps.
10. Note: Conducting random audits – a formal gemba walk – to check accuracy and effectiveness of work standards are discussed in the chapter on quality.

Developing good standard work: Work procedures are best developed with the involvement of the production operators. Hence, in the procedure described above, the revision and improvements to the initial draft standards must always involve the production operator. Otherwise we will have work procedures and documents that are different from practice. This discrepancy is often discovered during process audits. Where possible it is important to use FMEA techniques to review, highlight, and prevent workmanship failures. An example is step 8 in Fig. 3-1b; in this case it was discovered that the sequence of driving screws in to an assembly is important. Typically, all such multiple screw-in procedures require the engineer to initially determine if the screw-in sequence in important. That's why good and accurate work procedures are the basis for consistent and high quality.

A good (or even a bad) standard will become the platform for kaizen or continuous improvement of processes within a factory. Furthermore, the new and revised standards are the last stage of the PDCA improvement cycle. This is discussed further in the chapter on *quality*.

Poor standard work or lack of training: The impact of poorly written standard work or insufficient training can be disastrous. Fig. 3-4[14] illustrates a real case of what happened when production volume picked up and new operators had to be hired. The new operators were inadequately trained or were unaware of informal assembly procedures; that is, the improved work had not been standardized and documented. The effect, in this case, was that the results of continuous quality improvement were destroyed due to inadequate standard work and training or poor continuation of current standard work.

The same decline in quality can take place even when production volumes are constant but when there is employee attrition requiring new operators who are inadequately trained or current standard work is inadequate.

High Performance: Standard Work 39

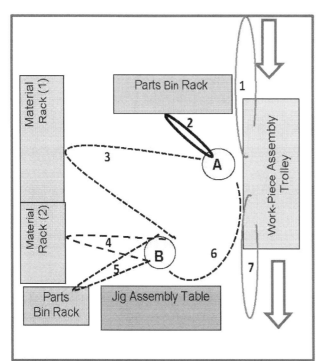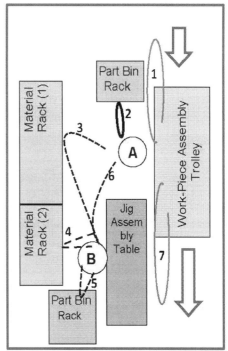

Station P 04: Travel and motion time calculation from spaghetti chart							
Travel Segment		Distance (m)	Times travel	Original Travel (m)	Original Time (s)	Revised Travel (m)	Revised Time (s)
1	Pull in workpiece in trolly	4	1	4	2	2	1
2	Get parts from bin	1.5	1	1.5	0.75	0.75	0.38
3	Get parts and sub-assembly from rack and place on assy table - 3x.	4	3	12	6	3	4.5
4	Get parts and sub-assy from rack and place on assembly table - 4x.	3	4	12	6	1.5	3
5	Get parts from bin	4	1	4	2	1	0.5
6	After assembly move sub-assy to main workpiece	2	1	2	1	1.5	0.75
7	After completing job move workpiece on trolly to next cell.	3	1	3	1.5	3	1.5
Total				38.50	19.25	12.75	11.63

Figure 3-3: Station layout, spaghetti chart, and calculation of travel and motion time

The Ohno Circle

According to Teruyuki Minoura[15], who worked under Taiichi Ohno the originator of TPS, Mr. Ohno used to order his trainees to go on the production floor and stand for hours within a circle he had drawn; he would then require his trainee to stand and observe the surrounding processes. According to Mr. Ohno, staring at a situation long enough will always uncover practices that are useless or help discover work and movements that create problems. By requiring this exercise, Mr. Ohno was communicating the power of deep and detailed observation of a manufacturing process coupled with thinking and analysis to come up with a better way of doing things. Indeed, Mr. Ohno is reputed to have told his production staff:

"If standard work does not change for one month you are salary thieves."

His point: Standard work and processes must be continuously improved. In fact we can verify that wherever we go in any factory floor and stand and observe, we discover wrong and right methods and come up with many ways to improve. Hence we too require engineers and managers to take this approach. Recently, on a typical day, here are some things we observed on a production line:

- A production operator deviated from the procedures because it was not possible to operate a machine per the standard procedure: It worked when the engineer did several dry runs – but the machine was not able to work smoothly in a repetitive production process. The engineer had been informed but had not fixed the process; meanwhile the operator had resorted to a manual process as it was faster and easier.

- An operator was soldering connectors on PC board assemblies with a robotic machine; after the process was completed, he cleaned the assemblies with a chemical and brush. We noticed that the standard work posted above him only required him to operate the robotic equipment on the assemblies and not to clean. So why was he cleaning the assemblies? They looked bad, so he was instructed by an engineer to clean them with a chemical wash. Later, it turned out the cleaning process spread contaminants on the assembly and a major rework was required for a few thousand assemblies resulting in shipment delays. Again a case of poor standard work, lack of process knowledge, and little training – we refer to the engineer not the operator, who only followed instructions.

Hence, it is imperative that in today's fast paced manufacturing environment, engineering professionals must take the proper approach to reduce waste and produce good standard work with the requisite training.

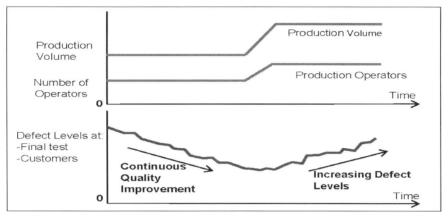

Figure 3-4 The effect of poor Standard Work on defect levels
This effect occurred when new operators were hired for new production lines; however, there was poor Standard Work available as a reference to train new operators. The same deterioration in performance can occur when production volumes are constant but if there is worker attrition and replacements are not adequately trained per the specific Standard Work.

Fig. 3-4 illustrates why many companies fail to hold improvement gains or see deteriorating quality. It may be obvious, but getting the organization or factory to ensure good documented standard work is challenging and takes both time and effort. Hence, what is required is "living" standard work – standard work that changes as the process changes and improves and operators are continuously trained in the new standards.

Changing standard work when production models, options, or configurations change: When the production line switches to a new model or option, the line may require to be re-configured and the standard work documentation for the next production run must be used:

- To inform the operator of the new work standard for the different model or option.
- To check that the line is set up correctly.
- To check that required inventory of parts, equipment, and personal work requirements are correct.
- To inform the engineer and supervisor of the change so that they mange to the new standard.

Typically on a production line with high mix of products or configuration changes, the operator will already be trained to be multi-skilled and hence be trained and certified to work on several different workstations and products. Each work station will have a manual of several work standards for each different model or option. Nevertheless, it is crucial that the standard work instructions are communicated quickly, with a minimum of disruption. In such a case, one of the ways to inform operators of the product or option change is to trigger the entire production line via a model change procedure. Here is how it will work for a production line with multiple workstations:

1. The production supervisor will notify the line supervisor or leader of the impending change of model or option via a new work order change.
2. The line supervisor or leader will place a special model change box informing the entire line of the model change and details of the new model or option.
3. The first unit of the new model will flow down the line, either after or in the special box, to trigger operators to look at the new work instructions and to review changes in assembly or parts for the new model or option.
4. The *model change box* is placed to follow the last product of the previous production run.
5. The *model change box* announces the product change, new model number, and informs the operator of the model/option change and requests implementation the new work standard.
6. At this point, the operator can switch to the appropriate work standard on the computer display or bring up the appropriate printed documents.
7. In Fig. 3-5, we show an example of the procedure for informing operators of model changes. A special yellow-colored bin is used to trigger operators and the production line of special configuration requirements for the next model coming down the line. Here the first four models of the new configuration are rolling down the line in the bin. Initially each operator works on four work-pieces, although the line runs with single piece manufacturing. This is to communicate clearly the change by making the operator work on 4 units with the new configuration; the line will revert back to single piece production after this alert. The product in this case is a customized PDA running in small volumes.

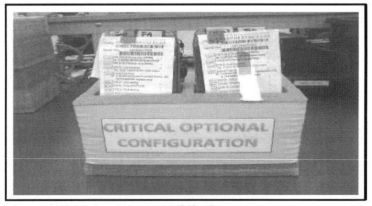

Figure 3-5: Model Change Notification

A special bin is used to trigger operators and the production line of special configuration requirements for the next model coming down the line. Here the first model of the new configuration is rolling down the line in a bin. In this case the product is a PDA requiring small volumes and special configuration.

The Training Process

Standard work is important and must be taught through an effective training process. Operator training must include generic skills-training and specific work standards training. Generic training includes practice of the 5S system, use of tools and equipment, and understanding how defects are created, detected, and prevented. This is followed by learning to read and understand work standards. For a typical operator, he or she must be trained to be multi-skilled and efficient with a series of selected standard work. Job-rotation is essential to keep the operator good at multi-skilled operations and to prevent quality fatigue.

In addition to honing the operator's skill through job-rotation, there are other benefits: The operator is less bored if she does different activities; it motivates the operator if she gets to do both simple and more difficult jobs; and it improves trust between operator and supervisor.

Training is best conducted by experienced supervisors or trainers with production experience; these trainers must be trained to conduct training and ensure operators understand the job at hand.

The training process should follow the PDCA (Plan, Do, Check, Act) cycle. A very good training process was publicized by the US War Production Board, via the Bureau of Training within Industry (BTWI), during the Second World War. Because of the extensive training required for new employees at factories during the war effort, the Bureau was able to establish an overall process of training, certifying superior trainers, and ensuring effective results. Unfortunately the lessons of history and the Bureau's success are not well publicized today. The BTWI recommended several areas of importance[16], shown below; we have paraphrased the original document and added additional comments.

1. **Get Ready to Instruct (Plan):** The key steps include:
    a. Have a timetable, with schedule and work methods that will be required
    b. Break down the job into important steps: This is essential for easy to understand and effective standard work.
    c. Have tools and equipment ready: This should be reflected in the standard work and should include all material, supplies, and parts required.
    d. Arrange the work station exactly as will be required by the operator when he or she starts actual production work. Hence the workstation is replicated during training and the basic 5S principles taught during generic training are reinforced here. This is the ideal requirement but usually only possible on the actual production line, hence on-the-job training is the best option.
2. **Instruction Guidelines (Do and Check):**
    a. Prepare the operator by putting him at ease. It is important that the operator is correctly attired per factory and standard work requirement and all safety regulations are communicated.

      b. Present the operation by showing the operator each job step at a time. This must be done clearly and deliberately.

      c. Get the operator to try out the work sequence slowly and accurately. Keep repeating until the sequence is mastered. The focus here is repeatability of the process, with an accurate and high quality output.

      d. Follow up by putting the operator on the job; make sure she has a coach to go to for help, check frequently until convinced the operator is doing it right. Typically the coach is an on the line trainer or the supervisor; both must be very skilled in the job at hand.

3. **Operator certification (Act):**

An operator will be fit to run one or more operations only after she is trained on the job, undergoes probation to fully absorb the job content, and is finally tested and certified. Depending on the complexity of the process the breakdown of such training should be: Initial on-job-training (1-5 days), probation (10-25 days), and test and certification. In addition multi-skill training and job rotation to sustain job interest are important.

Summary: Standard Work

Standard Work is an essential ingredient of manufacturing. It will include all the steps in the work sequence, the cycle time required to complete the work, and the inventory required at the workstation. These standards are essential for the simplest to the most complex job in the factory. If the factory can convert all work to documented and standardized work and ensure operators follow such standards, then work output can be of consistent quality. These standards then form the foundation for further improvement or kaizen. The ability to maintain good standard work, provide training in them, and ensure continuous improvement is the basis for high performance work and operational excellence.

    There is no easy process or silver bullet for completing standard work – it is different for every organization and in every area of work; but it must be done well. The key to standard work is keeping it clear and simple, so our operators can understand it and do their work accurately with high quality.

    One of the ingredients of standard work is cycle time; that is the topic of the next chapter.

Chapter 4

Efficiency: Manage Cycle Time

Time waste differs from material waste in that there can be no salvage.
The easiest of all wastes and the hardest to correct is the waste of time, because wasted time does not litter the floor like wasted material.
Henry Ford

Overview

Every assembly process on the factory floor takes time – whether it is done by manual assembly or by a machine. All the disparate processes in the factory need to be managed via the cycle time at each operation.

Cycle time is the second component of standard work. If cycle times are not well managed, we can have one workstation producing too much and another workstation producing too little. This will result in congestion and bottlenecks along the assembly line.

What we must do is to manage the production floor like an orchestra with every instrument coordinated, with no missing or superfluous notes, and with the music flowing exactly to the beat of the conductor. As with an orchestra we use the concept of takt time to manage the flow. A well orchestrated takt time system will ensure that the production line flows smoothly.

In this chapter we will discuss the concept and application of cycle time. Specifically, we will review:

- Takt time, cycle time, and total cycle time.
- Managing the factory floor via a takt time system to ensure smooth production flow.
- Maximizing efficiency on the production line by reducing constraints.
- Leveled production: Discuss the why and how of leveled production.

Takt Time, Cycle Time, and Total Cycle Time

Takt is a German word referring to the rhythm or beat of music. This was adopted in German manufacturing theory as *taktzeit*, which translates as cycle time. Takt time aims to match the pace of production with customer's demand; thus in manufacturing, takt time sets the pace for production. Hence, the time needed to complete work at each workstation has to meet the takt time goal.

Managing *taktzeit* in manufacturing also strives to remove the weakest link or slowest time in a series of manufacturing steps, in order to remove bottlenecks and speedup production; this is an ongoing cycle of improvement. In fact, the concept of *taktzeit* and the weakest link is the basis of the theory of constraints (TOC), which suggests that organizations must remove constraints to be more profitable. We will touch on TOC later in the text.

Typically the takt time sets the pace for a production line; this is sometimes called the pacemaker process. For example, in the manufacture of cell-phones, phones move along the assembly line at a fixed pace – this is the takt time. The formula for takt time will illustrate it better:

$$\text{Takt Time, } T_t = \frac{T_a \text{ (Available production time in a day or net available time)}}{D \text{ (Customer demand or units required per day)}}$$

Where
- T_t = Takt time
- T_a = Net available production time in a day (secs or minutes).
- D = Customer demand or units required per day.
- Note: T_t and T_a must use the same unit of time (secs or minutes).

Let's look at an example:
During a 12 hour shift, gross time is 12 x 60 = 720 mins.
Lunch break is 60 mins, 3 breaks take a total of 45 mins, and 15 mins are used for a daily clean-up and production meeting.

Therefore, T_a, available time, is 720 − 60 − 45−15 = 600 minutes.
Assume that customer demand, D, is 600 units a day.
Therefore takt time = T_a/D = 600/600 =1 minute.
Thus products, in these case cell-phones, must move along the production line at 1 minute intervals, and also exit the line at one minute intervals.

Takt time Vs total cycle time: However, the takt time is not the time taken to build the product; this will be much longer. If there are 50 stations on the assembly line, the time taken to build the product, or **Total Cycle Time,** is 50 x 1 minutes or 50 minutes; but the total cycle time is even longer if we include sub-assemblies that are purchased or built elsewhere.

Managing Takt Time on Sub-Assemblies: What about the takt time from feeder lines or subassemblies that are built elsewhere in the factory? Ideally, the sub-assembly line should run at the same takt time or pace, so it delivers these assemblies at the same pace to the main line. This would ensure that there is no accumulation of inventory between the final production and sub-assembly lines. But this may not be the best option. We could have the sub-assembly line delivering different parts to several final production lines; this would mean it builds for one line than for another, then back to building for the first line, creating a buffer WIP inventory.

Why Manage the Factory Floor via a Takt Time System?

Planned and predictable output: When all stations along an assembly line run at the same takt time, the product under assembly moves along smoothly at a pre-determined rate giving a predictable output from the line. Even if the product is assembled in production cells, this benefit can be achieved. The final assembly line also sets the pace for the rest of the factory. This is often called the pacemaker process as it sets the pace for all the upstream processes or the supporting sub-assembly lines, cells, or suppliers that must feed the final assembly line.

Repetitive efficiency and quality: All workstations along the assembly line will have a work standard and a trained operator. Hence a production line that builds the same product again can use the same operator or a multi-tasked operator and get repetitive good results.

Visibility of quality hiccups or bottlenecks: Let's suppose a production line is running smoothly and delivering predictable output; then if a bottleneck occurs, it will be immediately noticed. When a hiccup occurs (due to an operator having a problem or if products are failing at a test station) at any point on the line, there will be a traffic jam and inventory will build up just before the problem station. Therefore, the problem can be quickly identified and fixed.

Unbalanced lines (i.e. workstations running at different cycle times) will cause bottlenecks, delays, and inventory buildup. When workstations on a line are unbalanced, production throughput will drop. A review of workstation cycle times and improvement is required. However, when a production line is balanced, there will be a minimum of inventory and high equipment utilization. We will discuss this further when we review *Little's Law* and continuous flow.

Takt Time Vs Cycle Time

We use takt time to define and mange our overall assembly process; it basically gives us the timing for workstations to meet customer or planned demand. However each individual workstation in the assembly process will have a cycle time Tc: Where Tc = cycle time = the time taken to complete work per standard work at a workstation.

Cycle time will vary at each work station but *it is always planned to be less than the takt time* of an assembly process. Look at Fig. 4-1. The figure shows the takt time for an assembly process targeted at 60 seconds. Per our previous calculation this means that the demand for this product is 600 units per day. However, the various workstations on the line all have cycle times less than the takt time. Therefore: Tc ≤ Tt. In such a case this line will run smoothly, assuming there are no operator, equipment, or quality issues.

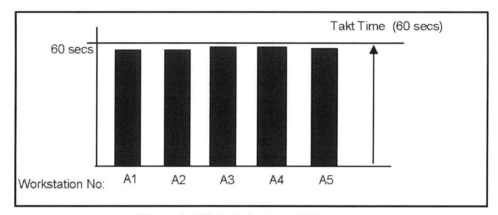

Figure 4-1: Takt time Vs cycle time

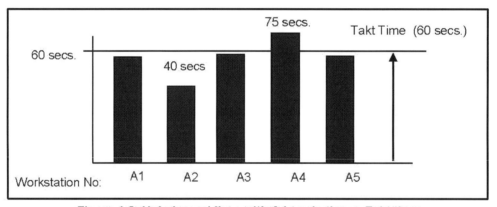

Figure 4-2: Unbalanced line, with A4 cycle time > Takt time

Takt time is basically a customer parameter and is driven by customer demand or forecast for the product. However on the assembly line and operation floor, the production lines, workstations, and equipment are measured by cycle times.

Definition of cycle time at a workstation: Let's look at the definition of cycle time:
Cycle Time = arrival time +queue time +setup time +process time +batch time.

Where:
- **Arrival time** is the time to move a job from the previous station and includes walking time
- **Queue time** is the time the job is waiting to be processed or to move to the next station (after processing)

- **Setup time** is the time it takes to set up the equipment for doing the job
- **Process time** is the actual time to process the job at the current workstation
- **Batch time** is the average waiting time to batch the job into a quantity of more than one unit. This would be zero for one-piece operation.

Looking at the cycle time definition, it is very clear that the process time is the only productive time, while everything else is waste or *muda*. Hence in preparing standard work, the role of the engineer is to minimize muda.

Challenges with the Takt Time and Cycle Time System

We have mentioned the benefits of the takt time system. Nevertheless, there are issues and challenges that come with the takt time system. These include:

- Selecting number of stations and operators.
- Cycle times on production lines.
- Re-balancing cycle times between workstations on a production line to ensure a smooth flow of production.
- Managing bottlenecks due to material and quality hiccups.
- Improving efficiency by reducing cycle times at each station and reducing total cycle time.

We will discuss these issues and challenges next.

Selecting number of stations and operators: Let's look at an example. If the takt time for assembling a digital-clock is 60 secs and the total cycle time (time taken to assemble the clock) is 300 seconds, then the number of stations required will be:

$$\text{Number of operators or workstations} = \frac{\text{Total Cycle Time}}{\text{Takt time}} = \frac{300}{60} = 5$$

After the production line is setup and running smoothly, it will look like Fig. 4-1.

Takt times on production lines: Takt times for consumer products may range from 5 secs to 300 secs (5 mins). Complex products may take longer. A station cycle time beyond about 5 mins may cause the operator to get bored or lose focus; often this results in *quality fatigue* causing assembly or quality defects. Remedies for this include job rotation of multi-skilled operators and frequent breaks.

Pacing production lines: Production lines using a conveyor system often mark their conveyors with cycle time markings. Refer to Fig. 4-3. Here the operator picks a work piece from the conveyor as it comes in front of her, works on it, and places it back in the same section; then she waits for the next work piece and repeats the cycle. The speed of the conveyor determines the cycle time she needs to conform to. Any station that cannot meet the required cycle time needs attention and resolution.

Re-balancing cycle times: When a line is operating smoothly as in Fig. 4-1, we can calculate the output using the takt time formula:

$$\text{Demand (or output)} = \frac{\text{available time}}{\text{takt time}}$$

Therefore output = $\frac{36000 \text{ secs (in a 10 hour shift)}}{60}$ = 600 digital-clocks

Figure 4-3: Cycle time markings on a production line

But there may be design and assembly changes to the product, resulting in changes to the station cycle times. Look at figure 4-2: Some stations are far askew from the takt time.

As the products move from stations A1 to A5, there will be some imbalance. At station A2 the operator will finish her job in 40 secs and will be idle for about 20 secs until the next job arrives at her station. Meanwhile, the operator at station A4 has to work at a cycle time of 75 secs. In this case there will be a gradual pileup of jobs between stations A3 and A4, because operator A4 is taking more than the takt time to finish his job. This pileup will become a bottleneck and the capacity of the line will decrease because the speed of the line is limited by station A4, whose cycle time is 75 secs. The line cannot manufacture a product faster than 75 secs each, even though the desired takt time is 60 secs. Therefore, the effective takt time is 75 secs. Hence, if we use the takt time formula, with a takt time of 75 secs:

Output = $\frac{36000 \text{ secs (in a 10 hour shift)}}{75}$ = 480 digital-clocks

We now get an output of 480 clocks. Fig. 4-1 shows a well balanced line; while Fig. 4-2 is an unbalanced line, which is inefficient and needs to be fixed quickly by rebalancing.

Line Balancing

In Fig. 4-2 we show a situation where the line is unbalanced – this can happen for many reasons. Initially, when a new product is introduced, there may be imbalances and work procedures will be in a state of constant revision until the line is balanced. Imbalances can also happen as the production process changes, is modified, improved, or during a start up stage. We need a well balanced line to ensure maximum efficiency of operators and maximum output from the production line. The benefits of a well balanced line result in proper distribution of work to ensure that all operators will be equally occupied and there is no wasted labor.

Therefore it is essential that we minimize cycle time variation between work stations: Cycle time is a parameter that must be monitored and tracked to minimize inefficiency. There are several components of variation that need to be monitored and improved:

- *An imbalanced line* is easy to see but can be a challenge to fix. In such a line production will be in a stop-and-go phase which will be very unstable.
- *Variation between work stations leads to waste* for those stations that are running cycle times below the takt time and bottlenecks at stations that are running high cycle times or above the takt time.
- *Variation of overall manufacturing cycle time* of a product leads to unpredictability of production output and waste in the system. This will show up in unpredictable schedules, unplanned overtime, and work in process inventory buildups.
- *Unintended impact of reducing cycle time:* When you reduce cycle time at a workstation or machine, and if the station is an independent operation, has unlimited material, and no other changes are made on the line, the average WIP (work in process) on the line will increase and so will the assembly cycle time. In other words, if you have a line running in steady state and you reduce the cycle time through improvement or by getting a higher-speed machine, do not be surprised if both WIP and cycle time actually increase. We have seen this effect at the beginning of a paced production line and at machines that feed production lines. Refer to the chapter on continuous flow and Little's Law for more information. The solution is to quickly correct the output at the station or machine to minimize the problem.

Therefore it is necessary that we keep track of every production line or cell to ensure that we have balanced production lines. Poor line balancing can be easily detected by the keen eye of a supervisor or engineer.

Project: Line Balancing

Let's review an actual line balancing project[17] to illustrate the mechanism of the re-balancing process. During our previous discussion on preparation of standard work, we mentioned that after the first draft is completed the standard needs to be reviewed and improved thereafter. Typically some work standards will come out just right while others will be way-out due to difficulties with the process, miscalculation, or even design changes. Since it is essential to obtain maximum efficiency, we must review the work standards and improve where necessary.

Fig. 4-4 shows the cycle times of all operations in the production line of about 95 workstations. The line assembles commercial printers and the data was taken after running the line for a week. Note that the figure shows the cycle times for all stations pertaining to the manufactured product: Stations on the final assembly line and also those on the sub-assembly lines, which feed the final assembly line. In the ideal situation, we want all workstations in the factory building to the takt time. However, if the sub-assembly line feeds several lines, then the production pace will be different and additional equipment or buffer inventory will be required.

In Fig. 4-4, we can see clearly that the line is unbalanced – several high cycle time operations were limiting the production output. The highest cycle time operation was 60 seconds at station A42, hence this limited the system takt time to 60 seconds giving an output of 60 units per hour. Furthermore, congestion and high WIP was occurring on the line.

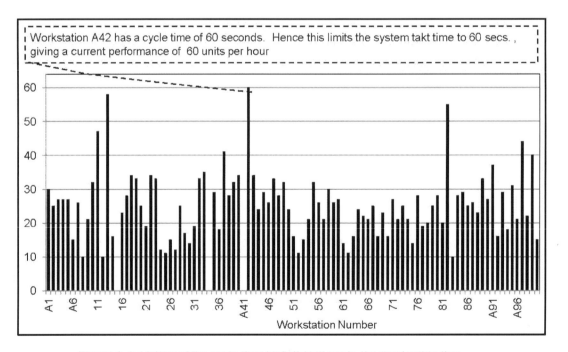

Figure 4-4: Listing of the cycle times of all stations in the production line

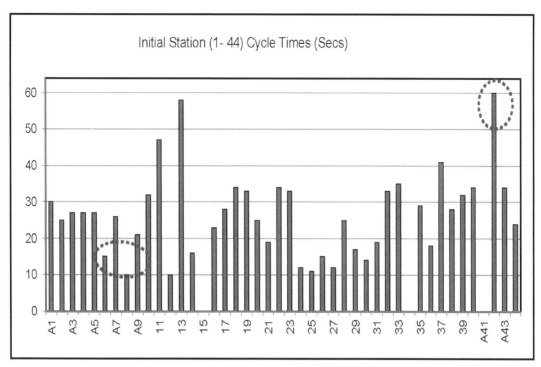

Figure 4-5: Reviewing and selecting areas workstations for improvement
In the figure we look at the first 44 workstations from Fig 4-4 and identify stations that need balancing. Workstations A6 and A8, and station A42. These stations must be analyzed with the objective of balancing them closer to the desired target cycle time.

The project goal was an output of 144 units per hour or 1500 units per 10.5 hour shift. This required a system takt time of 25 seconds and hence a maximum cycle time of 25 seconds at any one workstation.

Analysis and improvements: Since this project needed to move quickly, it was decided to make improvements via two improvement cycles: First targeting a more reachable 32 second cycle time followed by a second cycle with an aggressive 25 seconds cycle time. This project took two weeks for the first improvement and another two weeks for the second improvement. Production lines or cells with lesser operations should take a shorter time for improvement.

A detailed review of the first 44 stations is given in figure 4-5. Initial line output was 60 Units per Hour (UPH), limited by Station A42. In the figure we show the first 44 workstations from Fig 4-4 and identify and analyze two stations that need re-balancing. Workstations A6 and A8, and station A42. We had to analyze these stations, with the objective of redesigning the processes to move them towards the desired target of 32 seconds. These two stations represent the two extreme of cycle times: Very low and very high compared to our target.

Moving on to Fig. 4-6, we did an analysis of stations A6 and A8: both had low cycle times of 15 and 10 seconds respectively. Since the stations were adjacent it as possible to eliminate station A6 and merge its sequence of operations (SOO) with station A8. This resulted in an improved station cycle time at station A8 of 25 seconds, which was closer to the target. Next we look at Fig. 4-7 and station A42. This had a cycle time of 60 secs and the way forward was to split it into two stations. We achieved this by re-designing the process with stations 41 and 42. Note that station 41 did not exist before, but as we renumbered the stations after project completion, the same numbers were then used to identify the stations before the project started. After making these adjustments, the process for the stations A41 and A42 is shown at the bottom of Fig. 4-7.

The re-balancing process was carried out for all the critical stations until the team was satisfied. The outcome is shown in Fig 4-8. The performance after rebalancing improved: The maximum cycle time at any station was now 34 secs, which gave a production volume of 106 units per hour. This first improvement gave a productivity increase of 176% with a labor reduction of 10%. Refer to Table 4-1 for the analysis.

However, the team was still short of the goal of 144 units per hour. Looking at Fig. 4-8, it was obvious that there was more opportunity for improvement. There was still wide variation of cycle times between stations: There were about 15 stations running above 30 secs and about 12 stations running below 20 secs. Considerable waste remained.

A second round of improvement was conducted and a maximum cycle time of 27 secs or 133 units per hour was achieved (not shown here). The overall productivity increase for the project was recorded as: Units per hour increased from 60 to 133 or 221%, with reduction in operators of approximately 15%.

Measuring line efficiency: It is imperative to analyze and measure the efficiency of production lines or cells. This can be done by listing all workstations, recording their cycle times, targeting improvements, and finally measuring the improved process. We have shown this in Table 4-1, which is an extract from an Excel spreadsheet of the line we discussed in Fig. 4-4. The worksheet gives an analysis of the line efficiency before and after the first improvement cycle. In addition it highlights areas for further improvement.

Project conclusion: We can see clearly that after the line was re-balanced, there was better allocation of work to operators, bottlenecks were reduced, and the overall productivity increased.

Efficiency: Manage Cycle Time 55

Analysis and improvement plan for workstations A6 and A8:
- Station A8 CT (cycle time) = 10s; Station A6 CT =15s.
- Improvement plan is to eliminate station A6 and merge its sequence of operations (SOO) with station A8.

BEFORE: Station A6 (CT=15s) Station A8 (CT=10s)
Route sensor install pivot Installation of pick shaft

 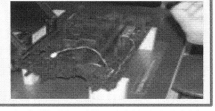

AFTER: Station A8 (after merging stations A6 into A8), CT =25s.

Figure 4-6: Analysis and improvement process

Analysis and improvement plan for workstation A39
- Station A42 (CT=60s) has 2 major sequence of operations.
- Improvement plan is to split this operation into 2 stations

Sequence 1
Prepare and install rollers

Sequence 2
Insert & install roller assembly

BEFORE: Station A42, CT = 60s.

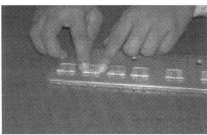 

AFTER: Station A41 (new), CT =32s   Station A42, CT = 29s

Figure 4-7: Analysis and improvement process

56   *Winning with Operational Excellence*

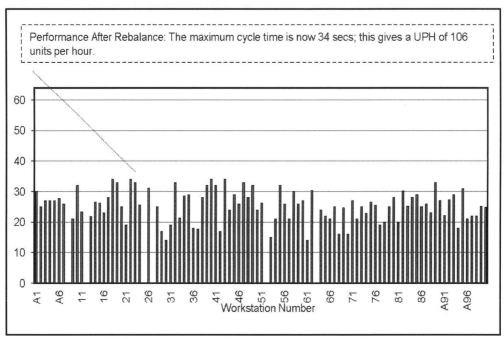

Figure 4-8: **Station Cycle Time After First Rebalance.** Compare this to Fig. 4-4.

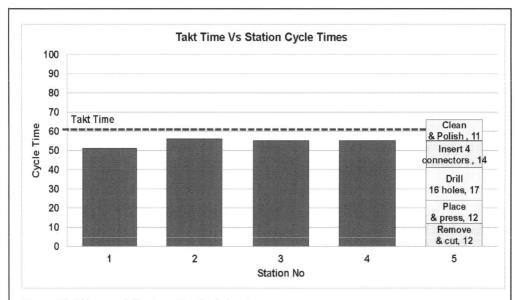

Figure 4-9: A Yamazumi Chart used for line balancing.
In this chart cycle times are displayed for 5 workstations in a cell. The cycle time for station 5 is segmented into the assembly steps for that station; in this case the cycle time for station 5 is exceeding the takt time. Typically, the assembly segments for all stations is shown, from which opportunities for improvement can be identified.

Yamazumi Charts

Yamazumi charts can be used to facilitate line balancing. A Yamazumi chart is a very useful tool for displaying cycle times and improving line balancing on a production line. Yamazumi is Japanese for heap, mound, or stack; hence a Yamazumi chart is a stacked bar chart, which shows cycle times at various stations Vs the takt time of the production line or cell. A Yamazumi chart is used visually to present the work content for a series of assembly operations. Typically, each stacked bar is segmented into the individual tasks at each work station. Fig. 4-9 shows a chart for a production cell.

Preparing a Yamazumi chart: The steps to construct a chart are:

1. Observe the tasks performed by the operator at the first assembly station.
2. Note the first task and the time taken for the task. This is the first segment of Yamaguchi stacked bar chart.
   - Refer to Fig. 4-9, assembly station 5: The task is "Remove and cut" (the detailed step is "Remove sheet metal from incoming & load into press and cut"), with a time of 12 seconds.
3. Next, observe the second task at the first assembly station.
   - Refer again to Fig 4-9, assembly station 5: The task is "Place and press" (the detailed step is "Move work piece into press and bend"), with a time of 12 seconds.
4. Repeat the sequence for all the tasks at the assembly station.
   - Refer to Fig. 4-9: The remaining tasks are "Drill 16 holes", "Insert 4 connectors", and "Clean & polish". All the individual tasks add up to the stacked bar at station 5. In this case the total cycle time for this station is 66 seconds, which exceeds the takt time.
5. Complete the chart by repeating this procedure for all the assembly stations.
   - Refer to Fig. 4-9 for the complete Yamazumi chart for the production cell; however it does not show the segments for the other stations.
6. Charts can be constructed manually on a white-board or on paper with an Excel program. Excel programs to construct charts are available – refer to the appendices for more information.

Displaying and using Yamazumi charts: Charts can be drawn on a white-board or printed on paper and displayed beside a production line or cell. As the production team improves the line, the charts can be updated real-time for discussion and further improvements. Many engineers use these charts in lieu of the work-combination chart, Fig. 3-2, which we discussed in Chapter 3. This visual display of individual tasks, cycle time, and takt time for an assembly line or cell is a very useful and valuable tool for line re-balancing and improvements.

Table 4-1: Extract of the workstation balancing sheet used to analyze efficiency

Balancing Worksheet			Product		Kite	Date		Owner	John. M.
Line	Station	Process	Cycle Time before improve	UPH	Cycle Time after improve	UPH	Capacity with 100% efficiency	Capacity with 90% efficiency	Remarks
Main Line	A1	Install S/N, Roller, & Pad	30	120	30	120	1260	1134	
Main Line	A2	Install Sp + Stop + Adjust	25	144	25	144	1512	1361	
Main Line	A3	Assemble ramp, Block, Pick, & install	27	133	27	133	1400	1260	
Main Line	A4	Install Pad, Shield, Guide, & Cover	27	133	27	133	1400	1260	
Sub Assy.	A5	Install Absorber, Eliminate, & Sensor	27	133	27	133	1400	1260	
Sub Assy.	A6	Sensor route & install Shaft	15	240					Remove A6 after merging into A8
Sub Assy.	A7	Prepare Assy. Pick Shaft and Roller	26	138	26	138	1454	1308	
Sub Assy.	A8	Installation of Assy. Pick Shaft	10	360	28	129	1360	1224	Merge A6 SOO here
Sub Assy.	A9	Install Pick Assy. & Rack	21	171	21	171	1800	1620	
Sub Assy.	A41	Prepare and Install Rollers			32	113	1181	1063	Split A42 into 2 stations: A42 & A41
Sub Assy.	A42	Insert & install roller assembly	60	60	29	124	1303	1173	
Main Line	A90	Power Up Testing	27	133	27	133	1400	1260	Six test stations, each at 160 sec
Highest Cycle Time and UPH			60	60	34	106	1111	1000	SUMMARY

Capacity is based on one shift of 10.5 hrs. The station with highest cycle time after the first improvement is not shown.

Managing Bottlenecks Due to Material and Quality Hiccups

Quality problems at any station can create bottlenecks, causing a line to stop. The options to resolve include:

Remove the failed product: The product can be removed from the production line to a rework station and production continues with the next unit.

Activate the jidoka/andon system in the factory: The production engineer is called to review and resolve immediately. This is discussed further in the section on quality.

Keeping safety stock in finished goods or sub-assemblies: These are used to compensate for any rejects that occur in the line. This will keep the production line moving in order to meet its target takt time.

Workstation Cycle Time Vs Overall Manufacturing Cycle Time

So far, our focus in this discussion has been workstation cycle time. However, overall manufacturing cycle time, which is the total time to manufacture a product, a notebook, or an automobile, is equally important. Managing and reducing overall manufacturing cycle time of a product is crucial for business success as it will reduce assembly costs, labor costs, and overheads. The result is cost reduction and a more competitive product. Here are some specific areas to focus upon.

Review effective cycle time: After a production line is balanced, it is appropriate to review production efficiency. One way is to determine the effective cycle time (ECT) of the production line.

$$\text{The effective cycle time, ECT} = \frac{\text{Operational cycle time}}{\text{Operational availability time}}$$

Let's look at an example: Let's say our target takt time is 60 seconds and we are planning to run at an operational cycle time of 55 secs (that is, we have managed to balance all our stations at 55 secs. or lower). However, we may have predicable or unplanned downtime due to machine or quality problems and can only run the line at 85% of the time. In this case our ECT = 55/0.85, which gives an ECT of 64.7 seconds, which is more than the takt time. Hence we will not be able to meet our production targets and will have to resort to overtime. Operational excellence requires that we have an operational availability of 95% - considered a World Class number – resulting in ECT of about 58 secs and meeting our production targets. To get high operational availability we need to ensure all equipment and machines are well maintained, production material does not run out on the line, and quality issues are rare and are managed by an effective jidoka system.

Analyze and review current situation: Understand the current situation. Use the PDCA improvement cycle to manage the process. Review the current situation by using the analysis shown in Table 4-1 to understand the individual situation at each workstation.

Reduce workstation cycle time: It is essential that we improve and find ways to reduce cycle time at each workstation via redesign of product and processes or better still by eliminating processes. As discussed in the previous chapter on standard work, in many work sequences there is non-work time, such as walking time, getting equipment or part time, and so on. With a good improvement effort these inefficiencies can be reduced. This will result in lower cycle times at many stations resulting in lower overall manufacturing cycle time. Waiting time must be analyzed: Is the operator waiting for the work piece or waiting for material?

- Motion time can be reduced if the operator's work area or cell is redesigned to reduce motion – refer to the use of spaghetti charts in the chapter on standard work and other ideas in the chapter on continuous flow.
- Work time can be reduced by redesigning the assembly process or providing better tools – e.g. press fit connectors Vs screwed in connectors, or auto tape dispenser Vs hand cutting tapes.

Cycle time variation: Another potential issue can be cycle time variation. A review and analysis of the process can help identify why there is variation: Variation due to work standards, difficult assembly process, or defective tools.

Reduce duplicate processes: Look at Table 4-1, station A90. There are 6 test stations operating in parallel because the test cycle time for each station is 160 seconds; hence 6 stations in parallel give a cycle time of 26 seconds. What are the opportunities to reduce test time and reduce the number of stations? Furthermore, there are often other duplicate stations because of the high assembly time; these can be reviewed for improvement or outsourced.

Outsourcing of complex assemblies: This is a viable option especially if the assembly requires a specific core competence that a supplier has, for example: Wire harnesses, complex printed circuit assembly, or a complete automobile door. Some manufacturers do not recommend outsourcing complex assemblies, as they feel that they must retain such core competencies in-house.

Should a line be running at 100% efficiency? After the line takt time is set and the line is running, the production output should be tracked and compared to production target. Results, deviations, and causes for deviation can be reviewed. Deviations or losses must be analyzed and resolved.

If there is 100% efficiency and the takt time goal is being met routinely, it may be a cause for suspicion. Why? *A 100% achievement rate may indicate that system cycle time may be much lower than takt time, hence a review of takt time is necessary.* There may be too much inventory or other waste in the line. This situation may also indicate that there are no rejects and all equipment is running perfectly, and that operational efficiency is 100%. This may be a cause for suspicion. In our experience if 95-98% efficiency is achieved it means that we are close to high efficiency even though some line stops are occurring: This requires review and improvement. Line stops will cause a state of tension, but this keeps the production team on their toes and minimizes complacency.

The Impact of Variability on Production Efficiency

Variability of processes can cause missed schedules, production delays, bottlenecks, inventory buildup, and productivity loss. Variability can occur in many factory processes: Supplier delivery, equipment downtime, machine setup time, machine capacity, incorrect standard work, lack of training, and unbalanced lines. There is a detailed discussion on minimizing variability in the chapter on continuous flow.

Leveled Production

For the takt time system to work efficiently the production demand needs to be stable for a reasonable period of time; this can be for a week, a month, or longer period. We do this by having a leveled production target, despite demand fluctuations. There are several types of demand leveling:

- Leveling the daily production demand
- Leveling the daily product mix within the daily demand. We will discuss this type of demand leveling in the chapter on Kanban.

Typically, we review the daily production demand or forecast and smoothen it out to provide a leveled production target for (say) an entire month. Some benefits and observations of leveled production are:

- There will be stable production and production lines will be balanced and run smoothly.
- There will be good labor utilization as the lines can stay balanced (through engineering effort) throughout the selected timeframe.
- There will be a minimum of adjustments and changes in production.
- This may result in excess finished goods inventory due to production target smoothening; however, this is acceptable due to the reduction in frequent line adjustments and production changes. As a result of leveling, the production line will be running efficiently.

Leveled Production Vs Build to Order

Ideally a production operation should be only building o order; this minimizes inventory in work in process and finished goods. Even so, a build to order environment will benefit from leveled production as the inventory buffer created will prevent the production lines from being jerked around by surprise orders.

Toyota Motor Company – with its World famous TPS – builds to order at its home base in Japan. Yet, it only does about half its business by build to order, the rest of production benefits from leveled production, resulting in some build to stock production. The combination however keeps both work in process and finished goods at a low level, and the production lines running smoothly.

Rebalancing Lines after Change in Demand or Forecast

Typically the takt time will be reset weekly, monthly, or at other interval based on the forecast and the revised leveled production targets. As takt time is changed, cycle time will be changed. This will impact line setup and management. If the demand drops by half, then the system takt time doubles. Let's look at or initial example at the beginning of this chapter – refer to Fig 4-10 below:

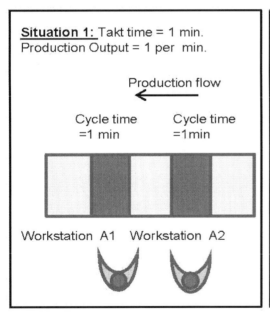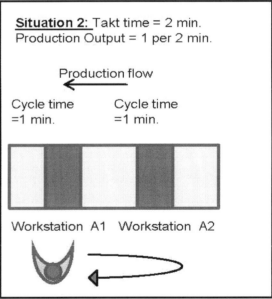

Figure 4-10: Impact of change (reduction) in takt time.
Operator doubles up in situation 2; hence the operators cycle time has doubled

Initial daily demand is 600 units for a 600 min day; therefore takt time is 1 min. With demand dropping by half to 300 units, takt time is now Tt = Ta/D, therefore Tt=600/300 = 2 minutes.

With the demand reduced, several options can be considered:
- The operator's cycle time has now changed, with a drop in demand the operator's cycle time has now increased or doubled. This is shown in Fig. 4-10. Note that the initial workstations are still in place; currently it is not advisable to merge stations A1 and A2 into a new work standard of two minute cycle time; doing so creates extra work in rewriting standard work, relaying out the line and so forth. Only with a permanent or drastic change in volume should such changes be made after all options are reviewed. Work reduced hours to manufacture less. If the factory has a two shift operation, reduce to a one shift operation and add overtime.
- Remove or de-commission lines, but this requires proper analysis.
- Reduce operators to decrease output on the same production lines.

We can reduce the number of operators and have each operator work on more operations – hence the operator needs to be multi-skilled and trained in various workstations across the factory.

If on the other hand demand increases dramatically, there are also many options, including overtime, starting a second shift, and adding more production lines.

If, however, the demand change is fractional and not a whole number, say +/− 30%, the solutions include overtime or shorter work hours with alternative work for operators.

Summary: Efficiency: Manage Cycle Time

A well managed takt time system to run the production line will give tremendous benefits. Takt time aims to match the pace of production with customer's demand; thus in manufacturing takt time sets the pace for production. The benefits include: A planned and predictable output, repetitive efficiency, and consistent quality. It will provide quick visibility of production bottlenecks caused by poor quality or inventory issues; this visibility is obvious from line stops, bottlenecks, or accumulation of defects.

When a production line is well balanced there will be a minimum of inventory in the line and there will be high equipment utilization. When it is not balanced, inventory will pile up along the line and reduce throughput. Therefore continuous improvement of cycle time and line rebalancing is also necessary. However, be warned that physically you can only go so far in reducing and rebalancing cycle times. A good rule of thumb is to balance workstations within ten percent of each other; beyond a point there will be diminishing returns and it will be better to focus on other areas that are more critical. We elaborate on this in the chapter on doing the right things or Hoshin Kanri planning.

The takt time system works best in a product environment that has leveled production. Therefore the factory needs a leveled production schedule, which is adjusted at appropriate intervals to smooth production flows and provide optimum workforce management. Leveled production will also keep production volume constant and reduce confusion and potential problems on the line.

A good takt time system will highlight inventory shortages or excesses. In the next chapter we discuss how to integrate inventory management with the takt time system via the kanban system and create smooth production flow and ensure just in time production.

Chapter 5

Just-In-Time Production: Kanban System

Just-in-Time means making only what is needed, when it is needed, and in the amount needed.
Toyota Motors Website

The aim of kanban is to make troubles come to the surface and link them to kaizen activity.
Taiichi Ohno

Overview

Every assembly process on the factory floor results in an accumulation of inventory – whether it is raw material, work in process (WIP), or finished goods. High inventory is considered the origin of many kinds of waste and must be reduced: Waste includes the space occupied on the factory floor, quality defects, and inventory cost. Inventory is the third component of standard work. Good standard work requires us to operate with a minimum of inventory at each work station. With lower inventory defects will quickly surface and inspection is easier because smaller quantities are easier to observe and check. The result will be higher quality products at lower cost.

Our objective is to implement just-in-time, or JIT, production and manufacture only what is needed, when it is needed, and in the amount needed. This will allow us to minimize inventory in the stockroom and factory floor. This means we must run the production line with material arriving just before it needs to be processed. We do this via the *kanban* system. Professor Yasuhiro Monden[18] describes *kanban* as a tool to achieve just-in-time production. The kanban system originated at Toyota Motors and has since been modified and improved.

In this chapter we will discuss the kanban system and review:
- The kanban system.
- Kanban system rules.
- Getting started with kanban
- Improving the kanban system
- Kanban system with suppliers

The Kanban System

The term Kanban means signal or card. In the original kanban system, the trigger to move a part came from a kanban (signal) card. In our context we will use it to mean both the card and the overall kanban system. However the kanban signal can also come from a signal card, an empty part-trolley, or empty kanban square.

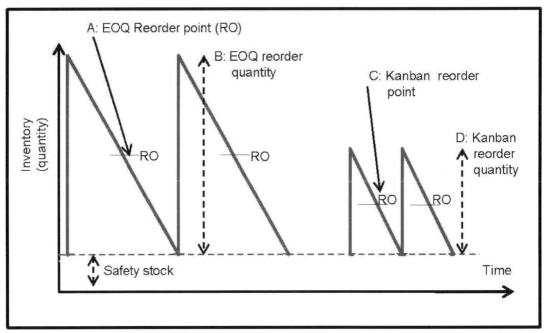

Figure 5-1: Inventory management via reorder point

Description of the Kanban Pull System

Kanban is a scheduling system that coordinates production and withdrawals to support just-in-time production. It works like the original re-order point method, based on the EOQ (economic order quantity) model. The EOQ model is based on a rule which looks at setup cost, fixed cost, and holding cost; the computed EOQ requires an order lead time that influences the reorder point.

The concepts of the EOQ system and kanban system are illustrated in Fig. 5-1. In the EOQ system:

The reorder point = Inventory needed to be consumed during the lead time before fresh parts arrive + safety stock.

At point A, a replenishment request, equal to the EOQ (point B), is triggered; fresh inventory arrives before the current inventory is exhausted, but there is safety stock, just-in-case it's late. This reorder process repeats ad infinitum. However, the EOQ system is more useful for purchasing and not for manufacturing; it is not helpful in multi-product manufacturing or small lot production; nor does it recognize the relation between quality and large lot production.

In the kanban system the emphasis is on just-in-time production by building smaller lots. The reorder quantity is influenced by the production lead time, which includes delivery transport time and machine setup time. If the production lead time is reduced, smaller lots can be manufactured quickly so that the reorder quantity is much less. This will result in a lower reorder point, shown at C in Fig. 5-1. Hence, the reorder quantity will be lower (point D). With further improvements in production lead time and more frequent withdrawals, the reorder quantity and safety stock can be lowered further. Therefore production lines can operate with less inventory and just-in-time production.

The kanban system is a pull system, i.e. material is pulled to the production line only when required. The mechanics of the traditional kanban system are illustrated in Fig. 5-2. Here downstream processes (like final assembly) will withdraw material when they need it from preceding, or upstream, processes. The upstream processes will then manufacture what has been withdrawn. This is the original two-card kanban system:

1. When material is received and placed beside the production line, the attached Kanban withdrawal cards are separated by the material handler (MH) and deposited in a withdrawal bin. Note that each pallet or container has one kanban card attached to it, while the container may have anywhere from (say) 1-50 parts on it.

2. When a number of pre-determined cards are deposited in the withdrawal bin, the MH will go to the buffer store of sub-assembly A with the withdrawal cards. He will take what he requires per the withdrawal cards. Examples of production and withdrawal cards are shown in Fig. 5-3.

3. For each pallet or container he takes, he will remove the attached production start card and attach the previously collected withdrawal cards to the new batch of parts.

4. The detached production start cards are deposited in the production start bins at the buffer store.

5. The MH takes the fresh material to the final assembly line and as the material is consumed, the withdrawal cards are again deposited in the withdrawal card bin.

6. Meantime, the production start bin contents are reviewed routinely and the production start bins are now collected and given to the production team to start production of new assemblies.

7. When the production units are completed as per the production start quantity, they will be stored in the buffer store with the accompanying production start card. In all cases, the physical units must travel with a withdrawal or production start card.

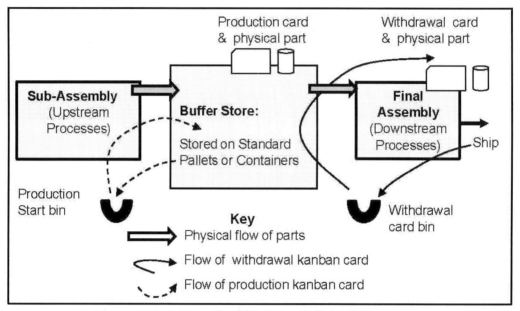

Figure 5-2: A Two Card Kanban Pull System

Withdrawal Kanban			
Location:			
Part Name:	Carriage Assembly	Previous Process	RM1: Harness & Mech. Assembly
Part No:	0540-1234		
Master Product:	RM 2300		
Carton Type	B-1	Next Process	Final Assembly
Carton Quantity	10		

Production Order Kanban			
Location:			
Part Name:	Carriage Assembly	Process	RM1: Harness & Mech. Assembly
Part No:	0540-1234		
Master Product:	RM 2300		
Order Quantity	10		

Figure 5-3: Kanban withdrawal & production cards

Kanban System Rules

For the kanban system to operate effectively it must follow rules. These rules are based on the combined experience of Shigeo Shingo[19] and our own. We give here several rules with comments and guidance; successful application will maximize the benefits of the kanban system.

1. **The subsequent or downstream processes should withdraw only the parts it needs from the preceding process.**
   - All withdrawals must be made only when needed and only with the accompanying kanban cards. It is critical that the kanban card accompanies the physical parts so that there is an audit trail.
   - The material handler (MH) who replenishes parts to the workstations is often called a *mizusumashi* or *water-beetle,* on Japanese production lines. A water beetle moves back and forth erratically on the water surface; this behavior personifies what the MH is doing all day – running back and forth on the factory floor.

2. **The preceding process can only manufacture the exact quantity withdrawn.**
   - This rule ensues that there is no overproduction by ensuring that production must match the quantity per the kanban production cards that are submitted. Yes, this can cause an issue and stress on the operators that are waiting for the signal to manufacture the next production order. Therefore, to ensure smooth operations, the upstream lines should be balanced and sized correctly for the current production plan; this requires leveled production in final assembly and throughout the operation.

3. **Kanban works best when the production schedule is leveled.**
   - We have already discussed leveled production. Basically the production schedule should be smoothened or leveled for a period of time. The kanban system can handle small production fluctuations by varying the number of cards going upstream. But large variations must be prevented through leveling. Therefore the leveling process is critical for both the upstream production processes and for suppliers, who need a predictable plan for a period of time.
   - Kanban works best in a repetitive manufacturing environment, where material and sub-assemblies flow at a continuous or steady pace. Kanban also works well in mixed model production, but to ensure timely manufacturing any machine used must be setup quickly as production converts from one model to the next. We will discuss quick setups and continuous flow manufacturing in later chapters.

4. **The products manufactured and sent to the next process must be 100% defect free.**
    - If defects are sent downstream, the line will have to stop production and the just-in-time process will breakdown.
    - Defects are less of a problem when an operation has high WIP as defects can be tolerated; but as the operation moves towards just-in-time production any defect will stop the line. Building defect-free products is addressed in detail in the chapter on *quality*.

5. **The kanban system must be continuously improved.**
    - The production lot sizes must be reduced by shortening production lead times: This requires reducing transport time and machine setup times. When machine setup is reduced the same quantity of parts can be manufactured more quickly and the kanban quantity reduced. Furthermore, reducing setup time improves machine utilization and provides quick changeovers from manufacturing one product to the next. We give more details on setup time improvement in the next chapter on machine management.
    - Fig. 5-4 gives a summary of the kanban improvement cycle. Every time there is a reduction of kanban cards or safety stock, the impact on production must be observed. Typically, we should do this in stages: Make a change, wait for the process to stabilize for a few days, then make another improvement and monitor the impact.
    - If there are line stops, the situation is analyzed for problems like:
        - An unbalanced line that is creating delays or roadblocks for material going downstream to the final assembly line.
        - High defect rate of assemblies, sub-assemblies, and other parts. Also, failures at test stations along the final assembly line.
        - Machine setup delays or breakdowns that are causing a delay in delivery from upstream lines.
    - Do note that there is a limit to reducing the number or kanban cards or reorder lot size: when the reorder lot size is too low, the frequency of pulls may get excessive; a high frequency of pulls requires hiring of more staff to deliver more often and also creates congestion within the factory. This must be taken into account when designing the system; a good balance between kanban quantities and delivery frequency has to be achieved.
    - As this improvement process continues, defects are reduced, and inventory in the factory will be reduced. With concurrent implementation of 5S and kanban systems, you will have a cleaner and leaner factory; and higher quality products.

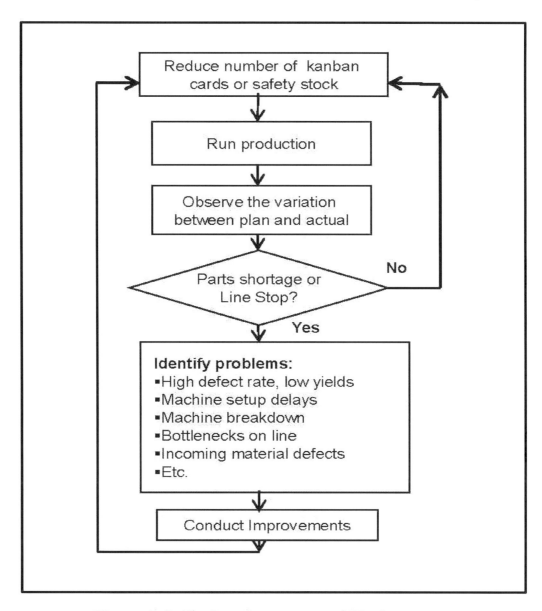

Figure 5-4: Kanban Improvement Cycle

One-card Kanban and Card-less Kanban Systems

In Fig 5-2, we have two storage points and a two-card kanban system. This is the original Toyota style system. In a large factory complex, with long delivery routes, proper documentation is necessary to keep control of all deliveries. For a smaller factory where downstream and upstream processes are closer, a one-card or card-less system will work:

- We can eliminate the buffer storage point and stock finished sub-assemblies at the end of the sub-assembly production line.
- In the final assembly line when sufficient withdrawal cards are accumulated, the MH will respond and proceed to the sub-assembly area, collect the required parts, and return with the required quantity of parts and withdrawal cards.
- The sub-assembly line responds to the lower finished sub-assembly stock and starts production again.
- Going to a one-card kanban system helps to streamline the system by reducing the number of kanban cards and reduces lost cards.

The one-card system can easily be converted to a card-less system, by using kanban racks or kanban squares to trigger the pull and production process. Figs. 5-5A and B show a Kanban Storage and Signal Rack System. In this case, the MH will respond to an empty rack or shelf. The empty rack is the signal to go to the designated sub-assembly line and withdraw required parts or alternatively to swop a full rack with an empty rack.

Figure 5-6 shows an empty kanban square to which the MH or *mizusumashi* must respond when she makes her rounds. In a well paced and well balanced line, both upstream and downstream lines are balanced and work proceeds with no downtime.

For small parts a two-bin kanban system is used: At the start of production both bins are full; when one bin is empty, that's the signal to pull parts and replenish the empty bin; meanwhile the production operator uses parts from the other bin. The process repeats throughout the day.

Reorder quantities for card-less kanban systems can be computed via the formula given in the adjacent box entitled: "How many kanban cards?" Typically, re-order quantity should be between one to a few hours of use. For small, inexpensive, components, it could be daily usage

Electronic kanban

Starting a kanban system with cards which are managed manually is a good way for operators to learn the system. Alternatively, going card-less is possible with kanban racks or squares. However, when dealing with operations in a sprawling factory, some sort of documentation is required. Within a large factory, quantity and delivery schedules can be documented on Excel spreadsheets or MRP systems. Going further there are electronic kanban, or E-kanban, systems, which can give real time visibility and trigger internal and external suppliers, including providing pre-set delivery schedules and quantities. However, we recommend running a manual kanban system, understanding its challenges and benefits; only then consider an E-kanban system.

Figure 5-5A (left) and 5-5B (right): Kanban Storage & Signal Racks.

In Fig. 5-5A, the operator at final assembly consumes sub-assemblies (assembled metal chassis) from the rack. One shelf is full, but the shelf below is empty – this is the *kanban signal* to the MH to get replacements. The MH will go to a rack in the upstream sub-assembly area (Fig. 5-5B) and bring it to final assembly. The present rack will be empty by then and it will be wheeled to the upstream sub-assembly line: This will be the signal to the sub-assembly line to build more units. The upstream line cannot build more than allowed by the empty rack.

In a well paced and well balanced line, both upstream and downstream lines are balanced and work proceeds with no downtime.

Figure 5-6: Kanban square

Typically there are two kanban squares marked on the production floor. At the beginning of production, both squares will have parts stocked in the squares. When one square is empty, it is the signal to the MH to get more parts, which will be placed in the empty square.

Meantime the operator will consume parts from the other square; when that is empty, it signals the MH to get more parts.

Getting Started With a Kanban System within the Factory

1. **Select parts for kanban system:** In this section, we only discuss the kanban system for use within the factory; running a kanban system with suppliers is discussed later. It's best to start at final assembly line. Typically the final assembly line will be manufacturing in one-piece production mode. Review the production process and select a number of sub-assemblies that are manufactured at upstream processes. Draw up a list of sub-assemblies that can be pulled by a kanban system; for example look at Table 5-1.

Table 5-1: Sub-Assemblies for Kanban		
	Sub-Assembly or Part	Source
1	Motor-Assembly	Line SB-1
2	Front Panel	Line SB-2
3	Main Chassis	Line SB-3
4	Wire Harness	Dock, Supplier delivered
5	PCB Assembly	Dock, From Block B

In this example we could initially start with the 3 sub-assemblies that are manufactured internally and defer the wire-harness and PCB Assembly which come from external operations or suppliers. A good rule of thumb is to use the Pareto principle and start with a small number of parts that represent the highest inventory cost.

2. **Decide on the kanban quantity required:** The kanban product quantity is determined from the equation given in the nearby box: "How many Kanban cards do you need?" Alternatively use kanban squares or racks.
3. **Select the container or pallet size to be used:** The withdrawal will take place in containers or pallets. Fixed quantities of parts will be placed in containers and withdrawn. This allows the kanban system to operate in a constant quantity withdrawal mode.
4. **Confirm the kanban withdrawal process:** There are several choices for triggering and making the withdrawal.
    a. **A fixed number of kanban withdrawal cards**: Refer to the box, "How many kanban cards". Here, a fixed number of cards must be predetermined; once this number accumulates from empty containers, the withdrawal must take place from the upstream buffer store. The predetermined quantity is basically the reorder point when the MH will go upstream to withdraw parts plus initiate a new production lot.

How Many Kanban Cards?

There are several guidelines for computing Kanban card quantity per part number. If you are just starting kanban pulls, the number of kanban cards, per part, can be computed from the original reorder point equation. We list one convenient equation here.

K= Required Kanban Quantity = (Production vol. X Production lead time) + Safety factor
     Kanban container capacity   Kanban container capacity

Or, $K = \dfrac{(P \times L) + S}{C}$

Where
K = number of kanban cards
P = production volume/hour
L = Production lead time (including processing & delivery time + setup time) in hours
S = Safety factor, use rule of thumb of 10% of (P x L). If production volume is leveled, S can be put as zero.
C = Container or pallet capacity

If hourly volume is 100 units, L=2 hours, C= 50, and we have leveled production, then

$K = \dfrac{(100 \times 2) + 0}{50} = \dfrac{200}{50} = 4$ kanban cards

Therefore, 4 kanban cards will accompany 4 containers.

Notes:

- It is important that the units are consistent: If volume D is per hour, L is in hours, while if volume D is per day, L must be converted into fractional days.

- The lower the machine setup time and delivery time, the lower will be the lead time L.

- The containers used should be standardized with a fixed capacity. *This allows the kanban system to operate in a standard quantity withdrawal mode.* The withdrawal frequency will change if the production target goes up or down the next month, but the line will still pull in standard quantities.

- The reorder point and the reorder quantity for the kanban system must also be set. For parts which do not require fixed lot orders, the standard withdrawal and production cards can be used. For parts that require orders in fixed-lot quantities (typical for parts requiring machine setup time), a signal kanban that is triggered by reorder point can be used.

- The goal is zero kanbans: In the kanban equation, if L and S = 0 and C=1, then regardless of demand, K or kanban cards = zero. This is the case when we have an assembly line moving parts one-piece at a time between workstations: This would be just-in-time production.

76   *Winning with Operational Excellence*

    b. **An empty kanban cart or kanban square**: This is an alternative to withdrawal cards and works well when the upstream line is close by. Refer to Figs 5-5 and 5-6 for more details.

    c. The above choices (4a and 4b) work well when the production lead time is low to zero, and the upstream line does not have to build a large lot of parts.

    d. **When the upstream line has a long production lead time**: A long lead time can be due to machine setup time or transport delivery time. In such a case, the final assembly line needs to trigger the upstream line at the reorder point; the reorder point will activate the upstream line to start building the next lot of material. A triangle kanban can be used for this process.

- Earlier we discussed the methodology of reorder point and reorder quantity. The triangle kanban uses this method. Refer to Fig. 5-7 and 5-8. The triangle kanban is sometimes called a signal card as it signals that the reorder point has been reached.

- Fig. 5-7 shows a triangle kanban signal card. Here the lot size (of 100) refers to the quantity that is ordered each time. The lot size is determined based on machine setup time and capacity; for example: a need to build multiple part numbers or just one part.

- The triangle kanban is fixed to the material so that when the box on the card is reached during consumption, the reorder point shown on the card is also reached; this is the signal for the reorder to take place.

- If the order goes to a machine, the time to pull the lot size = the machine setup time + processing time for the lot + scrap quantity (if any) + time to deliver the lot + a safety stock.

- If the order goes to the warehouse, then the time to pull and deliver plus the number of daily deliveries must be taken into account when planning the lot size. Per kanban theory: plan for a small lot size.

- Once the lot size is determined, the reorder point (which is 40 in this example) can be calculated. This will be the production build quantity during the time it takes to pull the lot size.

- At that point, the operator or MH can do one of several activities: Pull the triangle kanban card and take it to the upstream station and request to start production of the lot size, shown on the card. When the lot is ready, the triangle kanban is attached to the fresh lot and delivered to the final assembly line. Alternatively the MH can take the card to the warehouse to pull the parts. Delivery takes place before existing parts are exhausted. The triangle kanban is the preferred option when the production lead time is critical. Note: When multiple triangle kanban cards are involved, it is advisable to have a time stamp on the card to ensure processing priority.

- When the upstream line is building a high mix of different assemblies, the triangle kanban is often placed in a *heijunka* box or board and awaits its turn in the production queue.

Table 5-2: Batch Vs Mixed-Model Assembly							
Option 1: Batched production							
Day	Mon	Tues	Wed	Thurs	Fri	Sat	Sun
Black	1500	1500	1500	1500	1500	500	1500
Red	0	0	0	0	0	500	0
Green	0	0	0	0	0	500	0
Total	1500	1500	1500	1500	1500	1500	1500
Option 2: Mixed-model production							
Day	Mon	Tues	Wed	Thurs	Fri	Sat	Sun
Black	1300	1300	1300	1300	1400	1400	1500
Red	100	100	100	100	100	0	0
Green	100	100	100	100	0	100	0
Total	1500	1500	1500	1500	1500	1500	1500

e. **Use a heijunka box or process for mixed-model leveling:** Heijunka means "to smooth" in Japanese, and refers to the product leveling process that ensures a constant production rate in the upstream assembly line. We discussed two types of product leveling in the chapter on cycle time: leveling the daily production demand and leveling the daily product mix within the daily production demand. Let's do a quick review of the heijunka process.

- Automobile assembly lines do not build a large batch of one color or option and then switch to another color or option. The reason: Delivery for an order for a purple colored car may take forever, if the company builds a batch of white and black cars first. So instead of batching, they will do mixed-model production. Table 5-2 gives an example of batching Vs mixed-model production for colored telephones.

- Let' say the assembly line runs 24 hours/day, 7 days a week. It builds phones in 3 colors: Black is the high volume model, with smaller quantities of red and green. Its daily capacity is 1500 phones. The standard method (option 1) is to build in batches of one option followed by another option. In this case if the factory has daily shipments, it can only ship red or green phones every Saturday. However if the factory does mixed-model production as shown in option 2 and smoothes the daily demand at 1500, it can ship all colors everyday; thus minimizing lead times for customers while keeping inventories of slow moving red or green phones low.

- Smoothing of daily mixed-model demand can be done via a schedule worked out on a spreadsheet or manually, which is a simple method for mixed-model production.

- The manual heijunka process requires a white-board or a box that looks very much like option 2 of Table 5-2. It may have many pigeon holes, with the X-axis representing time and the Y-Axis representing the product in queue; each box could represent a one hour time slot. The Kanban cards are placed in the holes as per the downstream or customer order. Such manual scheduling is often used in restaurants; however it works equally well in the factory environment.

- If you have three *different* sub-assemblies required for the final assembly line, the heijunka-box can be used to visually prepare the hourly production plan. If you require 3 sub-assemblies, then instead of running them on separate days, you can run all three assemblies on the same day. So for example, you can run sub-assembly A for 7 hours, sub-assembly B for 2 hours and sub-assembly C for 1 hour, and then repeat the cycle. You can even run them repeatedly every day to minimize WIP and maximize daily shipments.

- Several triangle kanbans for different parts can be placed in the heijunka box or board. Each kanban represents the reorder quantity that must be built next. The reorder point takes into account the necessary queue time in the mixed-model line, so that the parts will be ready when required. Typically, the line is kept busy building mixed-models in small batches. This is preferable to building large lots for one model, causing the line to wait for other parts. Refer to the appendices for more information.

- Note that for the sub-assembly line to be able to switch from one model to the next, it requires quick set-up times for any equipment on the line, or quick change over time for assembly processes. We discuss quick setup times in the next chapter.

f. **Replenishment of raw material:** When replenishment of raw material is required on the line, a material signal kanban card can be used. This can be done via a material request kanban (see Fig. 5-7) that signals the reorder point, or alternatively via kanban carts or racks. Often the material kanban operates together with a triangle kanban: At the reorder point, the upstream process is triggered with the triangle kanban while the stockroom is triggered by the material kanban to deliver parts to the upstream process. However, when there is lots of low cost material on the line (screws, nuts, bolts, plastic parts, etc.) these can be delivered every shift by a simple Excel sheet tracker or visually triggered by empty kanban squares or an empty box in a dual container.

5. **Managing and preventing loss of kanban cards:** In a well managed kanban system, kanban cards seldom get lost. Nevertheless, kanban cards will get misplaced or lost; here are some precautions:

    a. To prevent loss or misplacement, the cards should not be too small. Although the kanban cards in Fig. 5-3 and 5-7 look small, it's best to keep them large and insert them in an a plastic cover to keep them from getting misplaced or lost.
    b. Review and replace dirty or soiled cards regularly.
    c. Make sure kanban bins are available close to the point of withdrawal.
    d. If the problem persists, do a study and observe the current process and look for limitations that need to be corrected.
    e. Kanban cards must always accompany a full container that has finished production or is travelling upstream.
    f. Once the first part is withdrawn from a full container, the withdrawal card must be pulled and taken to the kanban card bin. The MH needs to be aware that the production lead time clock is ticking, and he must proceed to the next process. The same urgency applies if a triangle or materials requisition kanban card is used.

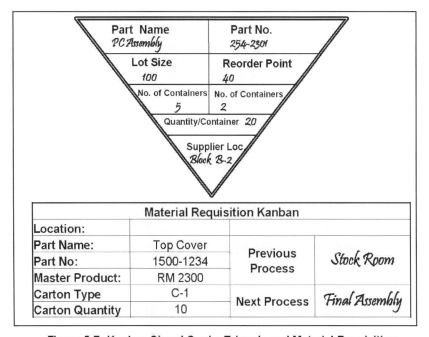

Figure 5-7: Kanban Signal Cards: Triangle and Material Requisition

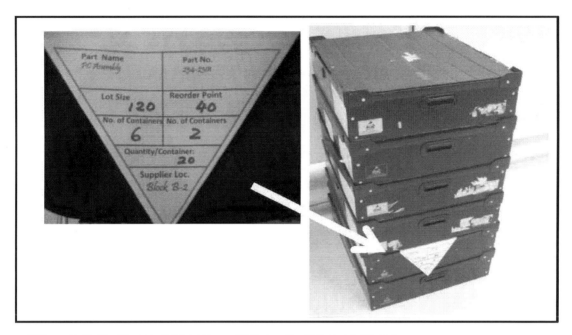

Figure 5-8: Use of Kanban Triangle Signal Card
The triangle kanban is sometimes called a signal card as it signals when the reorder point has been reached.

6. **Once the initial rollout of kanban is working**, we can rollout the system to more parts or another assembly line. Other points to note:
   a. The kanban system works best with a manufacturing philosophy of one piece production. Refer to the chapter on *continuous flow*.
   b. The production team will learn more about the details of the kanban system when it starts with a manual kanban system or a simple system documented on an Excel spreadsheet. When an E-kanban which is part of a large ERP system is used there can be confusion and pitfalls during the initial implementation.
   c. In Fig. 5-9, we show unedited photos of the first step of kanban implementation on a production line. Introducing kanban is often compared to draining water from a lake as rocks and other debris become visible. Similarly, reducing WIP inventory will surface other problems in an operation.

Implementing a Kanban System with Suppliers

When the kanban system is run within the factory, the normal mode of operation is the *constant quantity withdrawal mode*. With standard containers, the material pulls are in fixed quantities and if machines are used the material is built in fixed (small as possible) lot sizes. If demand varies during another month, the frequency of withdrawal changes, but the kanban card withdrawal quantities remain the same.

However, when implementing kanban with suppliers, the preferred method is the *constant cycle time withdrawal mode*. This is due to the suppliers being outside the factory, often far away. The suppliers have to ship product over long distances; hence freight costs as well as traffic congestion become an issue. Therefore as demand changes, the order quantity changes but the delivery process remains constant. Most MRP systems can be programmed to trigger kanban requests to suppliers. Alternatively, you can apply the same process that we have discussed for your suppliers in the approach we used for implementing kanban in the factory.

Figure 5-9: First step of kanban Implementation
The before and after photos above show improvements and inventory reduction after the kanban system is introduced. When water is drained from a lake, rocks and other debris show up; so in this case when inventory (boxes and pallets) is reduced, other issues/problems show up.
In the photo on the right, line rejects awaiting rework in the racks are now very visible and less than before. The inspection and test yields must be improved to reduce this problem. When this is accomplished, the racks can be eliminated or replaced with a small table. When the racks are eliminated, some other waste may show up behind the racks.

Major Challenge with JIT and Kanban System

We mentioned that the kanban system relies on a smooth leveled production plan. Still, there is the possibility that disruptions can occur due to equipment breakdown, product defects, or material availability. This will disrupt the JIT production process. All types of process variability will disrupt the Kanban system; we have a detailed discussion on managing variability in the chapter on continuous flow.

In a true JIT process, some variability can be compensated by having a production capacity buffer; this can come from overtime. However, if the production line is running full blast at a 24 x 7 pace, these will be very limited capacity buffer. Another choice is to have some finished goods inventory buffer to replace the lost capacity. But these are trade-offs that must be made for the benefit of JIT production and low WIP.

Summary: Just-In-Time Production and Kanban

The proven tool to support just-in-time production is the kanban system. Kanban is a scheduling system that coordinates production and withdrawals to ensure just-in-time production. We discussed the kanban system, including systems with cards and card-less systems. We also reviewed kanban rules and ways to improve kanban in an operation. Kanban works best in a repetitive manufacturing environment, where material and sub-assemblies flow at a continuous or steady pace.

We started this chapter with a quote from Taiichi Ohno on Kanban: *The aim of kanban is to make troubles come to the surface and link them to kaizen activity.* This epitomizes the very essence of kanban and just-in-time production. An effective kanban system, with its inherent (kaizen) improvement cycle, will lower inventory in WIP; with lower inventory defects will quickly surface and be resolved; consequently the operation will manufacture higher quality products.

Kanban also works well in mixed-model production, but to ensure timely manufacturing and small-lot production the machines used or feeder lines must be setup quickly as production converts from one model to the next. Therefore, one of the ways to ensure effective and proficient just-in-time manufacturing is to have quicker machine setups and more reliable machines. That is the topic in the next chapter.

Chapter 6

Machine Management: Quick Setup and TPM

The greatest task before civilization at present is to make machines what they ought to be: the slaves, instead of the masters of men.
Henry Ellis

Overview

Machines and equipment are used in every manufacturing process. They help to perform difficult tasks, increase efficiency, and improve quality of products. Machines, however, need to be managed to get results; specifically they must have quick setup times and high reliability.

In the last chapter we discussed how the kanban system can be improved by reducing production lead time, so that the production line can run with lower inventory and convert quickly from making one product to another. In addition, machines must provide predictable performance, deliver exactly what we expect of them, and have zero breakdowns.

In this chapter we will review several areas of machine management. Specifically:

- Reducing production lead times.
- Quick setup of machines and equipment.
- Total productive maintenance (TPM) of machines.
- Basic Requirement for TPM

Reducing Production Lead-time

There are several manufacturing strategies that help to reduce long production lead time, which is waste. Lead times can be reduced by doing one-piece production: This is discussed in the chapter on *continuous flow* and one-piece production.

Production lead time for a production process can also be reduced by reducing machine setup time and machine availability. Shorter setup times mean that machines can minimize the production lot size. For example, if the setup time for a machine is two hours, and if we used the standard economic order quantity (EOQ) equation we would run it for several hours. However, if we could reduce setup time to (say) one minute, we could run the machine for a few minutes and build a small lot size. This would help us to produce parts quicker and minimize inventory. This is the thinking behind the SMED program or Single Minute Exchange of Die. In addition, machine availability, via good maintenance planning, will help by ensuring machines run every time after quick setup.

The goal of the SMED program is to setup a machine within minutes. This will allow an operation to build products in smaller lots, run an effective kanban system, and facilitate production of a high product mix running at low volume for each product. Let's review how machine setup can be reduced. But first, refer to the box to see how Ferrari does its tire-change process.

How long does it take you to Change a Tire and Refuel?

If you had a punctured tire, it would take you anywhere from 15 to 30 minutes for one tire change. This seems reasonable. But if you were in a NASCAR or Formula One race, this would be too long. You would want the tire change at a pit-stop to take less than 20 seconds. In fact, at a Formula One race in the Singapore Grand Prix, we clocked the time to change four tires plus refuel a Ferrari at 5 seconds. What is the difference between you changing a punctured tire and the time taken by the Ferrari team? The table below highlights some of the differences:

Item	My or Your Punctured tire change process	Ferrari tire change at a Formula One pit-stop
Experience of personnel that make the change	None. Perhaps one tire change every 2 years	Trained and regular practice is provided.
Tool-Set	Basic tools somewhere in the trunk of car.	Special power tools, custom made wheels with only one nut to change per tire, etc.
Standardized work	None. Read the manual if you can find it.	Standard work improved with time and practice.
Performance measures	None. Fix it and move along.	Set target, which must be achieved.
Time taken	15 to 30 mins (for a tire)	5 Seconds (for 4 tires)

The difference is clearly due to the preparation, standard work, trained operators, and a good tool-set. Note: The Ferrari team had about 4 operators working on each tire, 3 on the refueling process, and 1 supervisor, giving a total of 20 operators. Nevertheless, what this shows is setup time can be reduced, if you have an aggressive goal and good modus operandi.

Quick Setup of Machines and Equipment

Machine setup time can be reduced by analyzing the basic steps of the setup procedure and improving in each step. The procedure is illustrated in Fig. 6-1 and discussed in detail below:

1. **Split setup time into external and internal activities**: External setup time refers to the activity that can be done while the machine is operating on another job. For example getting the tool or die from storage, preparing and cleaning the tools, making sure the tools are adjusted and ready for use or installation. Internal setup time refers to the activity that can only be done after stopping the machine. For example installing the tool or die or loading raw material or parts into the machine. Understanding and separating the two types of setup processes is necessary before we can proceed to reduce setup time.

2. **Move as many activities as possible to external setup:** Review all activity for setup and segregate into internal and external actions. After that plan to run all external activity and preparations before the machine is stopped for setup. This will include:
    a. Getting the tool, die, or other attachments from storage.
    b. Preparing and cleaning the tools.
    c. Making sure the tools are adjusted and ready for use.
    d. Getting raw material and parts for the next job
    e. Standardizing tool and die design to allow for quick change with a minimum of adjustments.
    f. Pre-loading the software program for the next job.

3. **Reduce internal setup time:** After transferring activity to external setup, look for opportunities to reduce internal setup time. For example:
    a. Reduce internal setup time by process simplification.
    b. Use quick disconnects and fasteners.
    c. Replace bolts with quick lock and unlock clamps.
    d. Eliminate need for tools.
    e. Minimize adjustments during setup.
    f. Standardizing tool dimensions to minimize adjustments when changing tools, dies, punches, etc.

4. **Reduce external setup time:** Even as you move internal setup to external setup, you must reduce the external set up procedures:
    a. Clean tools prior to initial storage so you only pull clean tools.
    b. Minimize adjustments when setting up tools.
    c. Improve and streamline processes to get raw material and parts.

5. **Eliminate the set up procedure**: This can be done only with proper design of products and tools, For example:
   a. Multiple cavity molds that make several different parts for a product.
   b. Multiple products that are designed to use similar parts; this is the basis for standardization and common parts usage in automobiles, TVs, and consumer products.
   c. SMT (for PC board assembly) machines can be setup with multiple kits of parts to run several different jobs in sequence.
6. **Train staff in the improved process; measure, analyze and improve performance:** After training the staff in the improved process, the next step is to analyze the current process to improve further. This often requires new ideas instead of tweaking the current process. Refer to the project illustrated in Fig. 6-2, where both breakthrough methods process and tweaking were done.

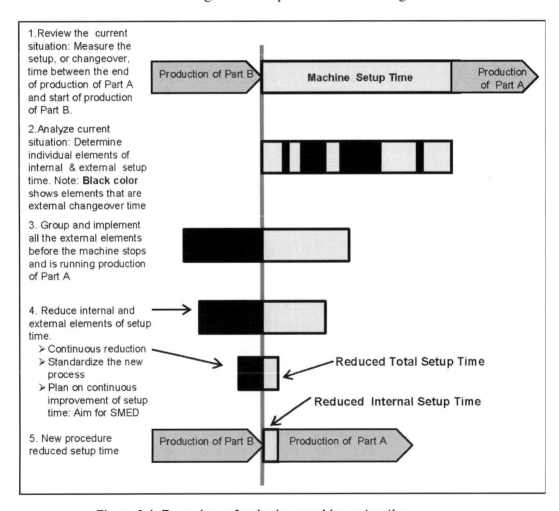

Figure 6-1: Procedure of reducing machine setup time

Project: Changeover and Setup Improvement

We have discussed methods of reducing setup time. The methodology applies to both individual machines, a set of machines running in sequence, or a production line. In Fig. 6-2, we illustrate a project summary[20] for reducing the *changeover time* in a printed circuit board (PCB) production line.

What is the difference between *changeover time* and *setup time*? Setup time is a subset of changeover time. The changeover time refers to the time required to switch a production line from manufacturing product X to product Y; the changeover procedure may require setup up of one or more machines, changing line processes, or even changing production operators. The setup time is simply the time taken to replace existing components, jigs, dies, or fixtures in order to manufacture a different product on a machine.

In this project, the PCB production line consists of two production lines running in sequence. Each production line consists of solder paste printing machine; a SMT machine to load components, chips, and ICs; and a reflow oven that solders the components onto the PC board. This line manufactures assemblies that are supplied to an automotive manufacturer in Japan and USA. Quality control and process control are critical on this line to ensure that the completed assemblies meet strict standards. Typically, critical automotive components and products are not allowed to be reworked if they fail test: They must be manufactured perfectly the first time, if not they are scrapped for reasons of reliability. This line has to run a high mix of different products requiring both high and low volume depending on demand; therefore it is imperative that setup times are low and quick changeovers help to meet market demand and fluctuations.

Fig. 6-2 provides a summary of the project. The original changeover time was 102 mins. Using the generic procedure discussed in Fig. 6-1, three improvement cycles were carried out. In the first improvement the external setup processes - which were software selection and component feeder preparation - were done prior to stopping the production line. However, the improved setup time of 78 mins was still unacceptable. The next step was to analyze the remaining internal setup times. After an analysis, countermeasures were proposed and two more improvement cycles were implemented. After project completion the setup time was reduced from 102 minutes to 20 minutes, which is a reduction in setup time of 80%. More opportunities remain. A goal of a few minutes – in line with SMED thinking is achievable - but that is the subject of another project.

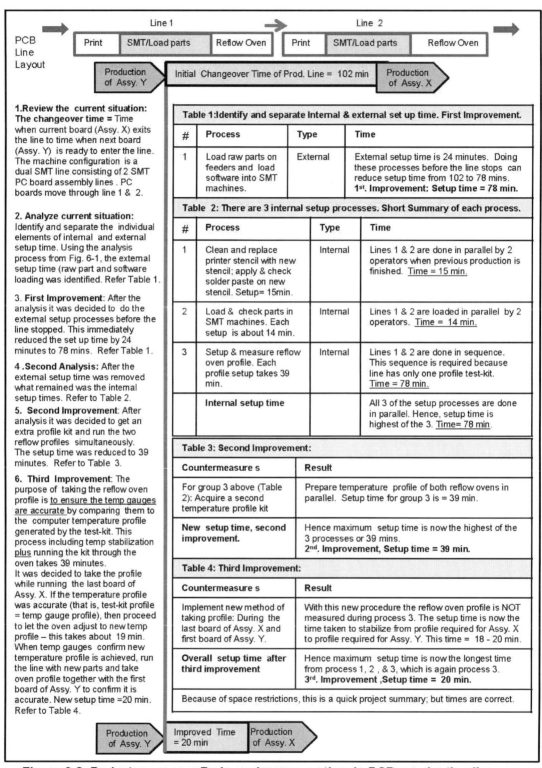

Figure 6-2: Project summary: Reduce changeover time in PCB production line.

Specific Benefits of Quick Setup Time

There are several manufacturing strategies that help to reduce long production lead time, which is waste. Moving parts and assemblies down a conveyor line decreases transport time; lead times can also be reduced by doing one-piece production in final and sub-assembly lines or in production cells.

Lead times can also be reduced by reducing machine setup time and availability. Shorter setup times mean that machines can minimize the production lot size. This is the thinking behind the SMED program or Single Minute Exchange of Die.

Let's look an example using the information from our changeover improvement project. Two companies: A and B manufacture printed circuit boards. Company A has a changeover time of 102 Minutes; Company B has a changeover time of 20 minutes. Each board takes an *average* of two minutes to flow down the line; that is the manufacturing time is two minutes. This is how they will perform if they both have to manufacture a small lot of 30 printed circuit assembly boards:

Company A:

- Company A will take 162 minutes to manufacture all the boards (changeover time + manufacturing time, or 102 + 60 = 162).
- First board will come out after 104 minutes
- Average manufacturing time for each board is 5.4 minutes (162/30=5.4).

Company B:

- Company B will take 80 minutes to manufacture all the boards (changeover time + manufacturing time, or 20 + 60 = 80)
- First board will come out after 22 minutes
- Average manufacturing time of each board is 2.7 minutes (80/30=2.7).
- If we look at the data, it is clear that Company B is more efficient than Company A.

Company B can:

- Build boards faster or build almost twice as many boards in the same time.
- Manufacture smaller lots quicker and more efficiently.
- Have better machine utilization, hence costs are lower.
- Detect any problems quicker due to faster feedback of issues.

Furthermore if Company B reduces the changeover time to meet a SMED goal of two minutes, it can build smaller lots quickly, more efficiently, and with lower costs

Total Productive Maintenance (TPM)

Not only must we reduce machine setup time, we must also ensure that machines work every time. For that we need total productive maintenance or TPM. The objective of TPM is to ensure that machines provide predictable performance, deliver exactly what we expect of them, with zero errors, breakdowns, or accidents; all this at a reasonable cost.

In the past the workforce was divided into operators who focused on producing parts and products, while skilled technicians were dedicated to maintaining the machines. However, with more complex machines, the maintenance workload had to be shared with operators; the alternative was adding more technicians. With TPM, skilled operators must do much of the basic maintenance. The more skilled-maintenance team must work on more complex machine maintenance, calibration, and managing overall equipment effectiveness (OEE).

A strong TPM plan requires us to:

- Understand the typical pattern of machine failure so that we are better prepared to minimize problems.
- Measure, monitor, and improve machine failure rate and overall productivity.
- Adopt a planned approach to preventing failures and maintaining machines: this includes training operators to run the machines more efficiently and to take daily care of machines so as to be the first line of defense in preventing potential problems. Technicians will attend to the more complex machine issues.
- Use technology to predict and prevent machine failures.

Typical Pattern of Machine Failure

Just like humans, machines can fail. Our objective then is to ensure that we maintain machines in a healthy condition for as long as possible. Machines that are well designed and maintained will have a failure rate that resembles a bathtub curve. The bath tub curve[21] shown in the Fig. 6-3 depicts the failure rate of machines over a period of time; however such a curve cannot be predicted or exact for a specific machine – the curve is a hypothetical projection. Typically a machine goes through three modes of failure: infant mortality or early failure rate, low failures during mid-life, and an increasing end-of-life failure rate.

1. **Early failures/Infant mortality:** This may occur in a new machine due to design issues or unreliable components. Manufacturers often do a burn-in to detect and filter out such issues before shipping the product to a customer. In a good pre-tested machine there will be no early failures – most manufacturers attempt to achieve this.

2. **Normal life:** During the normal machine life there should no or few failures. But random or wear and tear failures can occur as shown in Fig. 6-3. Such failures are difficult to predict or avoid. The goal of TPM is to minimize such failures.
3. **End-of-life failures:** In the long term all products will wear out or fail. Typically a few components will determine the end of life phase. Monitoring and managing critical, often mechanical, components can help to prolong end of life. Nevertheless, eventually repairs get too expensive or a repaired machine will not be able to deliver optimum performance.

Causes and Prevention of Machine Failures

The goal of TPM is to minimize failures during normal machine life. The following is a guide to causes of failure:

1. **Failure to maintain machines per a recommended procedure:** This is the most common cause and often due to cost controls and ignorance. Delaying maintenance can be expensive because it can lead to parts deterioration which can be expensive to replace. So it's pay now or pay later with lost production time and expensive machine repairs.
2. **Parts of a machine can deteriorate**: This is due to lack of regular cleaning or insufficient lubrication. Hence maintenance is a must.
3. **If a machine is stressed or overloaded, failure can occur:** This is similar to driving a car at very high engine speeds beyond the redline limit. Machines should not be driven beyond their specifications.
4. **Poor operator training**: As machines get more complex, operator training is crucial. Too often poorly trained operators run complex machines with inferior results and unnecessary rework.

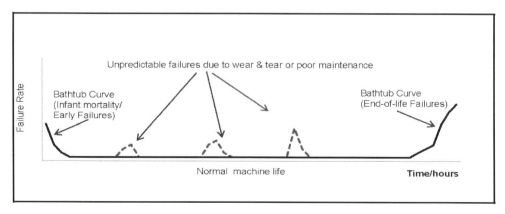

Figure 6-3: Bathtub curve - failure rate Vs time (operating hours)

What is The Failure Goal for Aircraft?

The obvious answer: Zero. Airplanes cannot afford any failures – causing flight delays or lives lost. Understanding how airplanes are maintained and how the prevent failures helps us to come up with better TPM strategies. Here are several strategies used for aircraft maintenance:

Routine and documented preventive maintenance programs: The manufacturer's MTBF part change schedule must be followed, no exceptions.

Aircraft maintenance checks: Aircraft maintenance checks are periodic checks that have to be done on all aircraft after a certain amount of time or usage. Airlines and other commercial operators of large aircraft follow a continuous inspection program approved by the Federal Aviation Administration (FAA) in the United States. The FAA issues a Continuous Airworthiness Maintenance Program (CAMP) under its Operations Specifications. The CAMP includes both routine and detailed inspections.

Airlines and airworthiness authorities casually refer to the detailed inspections as "checks", commonly one of the following: A check, B check, C check, or D check. Checks A and B are light checks, done monthly and three-monthly respectively. However, checks C and D are heavier checks. The D check occurs approximately every 4 – 5 years. This check is the most detailed and virtually takes the entire aircraft apart for inspection. (Source: Wikipedia)

Other Strategies and metrics used:

- Extensive use of software diagnostics while the plane is on the ground to forecast potential failures in parts or systems.
- Pilots (operators) use process quality assurance to monitor key processes and look out for potential problems:
- Power generated by each engine – is it balanced or within limits?
- Fuel consumed by each engine – is it balanced? If not there may be a potential problem.
- Internal cabin air-pressure – deviations from target may indicate leakages in the system.
- Instruments to provide radar warning of nearby aircraft or storm clouds.
- Instruments to provide warning of dangerous wind shear movements.
- Class failure information to predict and prevent future failures via checks, part replacements, or system repair. This information is typically obtained from other aircraft failures or Black Box data from recent accidents.
- Redesign of aircraft systems based on Black Box data or other failures.
- Grounding of aircraft for complete system check or repair if recommended by FAA.

Understanding these causes is the first step to coming out with an effective TPM plan. To get started we need to measure machine failure rate and their impact on productivity.

Measuring Machine Failure Rate and Productivity

In any operation, machine failure rates and productivity must be monitored and measured, with goals for improvement. There are several metrics for measuring failure rates, here are the key ones:

Machine failure rate = number of failures per time period = $\dfrac{\text{Number of failures}}{\text{Operating hours}}$

This can be stated as failures per week or month.

Mean time between failure (MTBF) = $\dfrac{\text{Operating hours}}{\text{Number of failures}}$

Overall equipment effectiveness (OEE) = A x P x Q

Where,

A = Overall equipment availability = $\dfrac{\text{Available time} - \text{downtime}}{\text{Available time}}$

P = Performance = $\dfrac{\text{Actual parts produced}}{\text{Theoretical machine capacity (often termed nameplate speed)}}$

Q = Quality = $\dfrac{\text{Production quantity} - \text{defective quantity}}{\text{Production quantity}}$

OEE is a holistic measure and goes beyond machine uptime and looks at availability, performance, and quality; *it's typically measured at machine level*. OEE measures several large losses, such as unplanned breakdowns, machine set up and changeovers, stops for any other reason, machine slowdowns, and defects or quality levels.

Each parameter (A, P, Q) is measured in %. Hence if each parameter is 90%, we get an OEE of 73% (from 0.9 x 0.9 x 0.9 = 0.73). Thus each parameter has to be high if we wish to achieve an OEE of 90%.

Basic Requirement for TPM

TPM goes beyond the standard preventive maintenance programs as it aims for predictable performance, with zero errors, breakdowns, or accidents. Furthermore it requires participation of operators, who form the first line of defense in ensuring that machines do not deteriorate prematurely. In addition the use of diagnostics and process quality measures is required as they will provide early warning of potential machine issues. Specifically we need:

1. **Goals of TPM.** The basic goals of TPM are maximum machine output with zero (to low) accidents, breakdowns, and defects.

2. **Operator and technician training.** The starting point of a good TPM plan is to provide appropriate and extensive skills-training. As machines get more complex, operator training becomes more crucial. Hence training and certifying operators to use machines is a basic necessity. Operator certification requires some theory, practical training, and basic cleaning and maintenance skills. Technician certification is equally important but needs to be more detailed. As discussed in the chapter on standard work, these operators and technicians must be multi-skilled: This is critical for a good TPM effort.

3. **Routine cleaning is done by operators and technicians.** A daily cleanup schedule for each machine should be documented and implemented. The daily schedule can be done by production operators and will include: Cleanup and removal of dirt, debris, dust, contamination, dropouts, and scrap. Both the machines and the surrounding areas must be cleaned.

    There must be a schedule for detailed cleaning of parts inside the machine: this can be done by a skilled operator or technician, who needs to open the machine and cleanout scrap and dirt. Figure 6-4 shows a dirty exhaust duct in a temperature-cycling test oven that was not cleaned for years. It only came to the attention of engineers when dirt and debris started showing up on finished parts that were treated in the machine. It reminds us of clogged arteries, due to poor diet and lifestyle. Hence, regular external and internal cleaning has to be part of the routine maintenance procedure.

4. **Prepare a schedule for routine lubrication of machines.** The lubrication procedures can be split into work done by operators and work done by technicians. Typically this is scheduled into weekly and monthly requirements plus may involve cleaning debris, oil leaks, and grease.

5. **Carry out routine overall inspection.** This activity goes beyond routine cleaning and requires operators to be trained to inspect specific parts of the machine. Determine category of inspections (loose stuff, damage, leaks, noise, excessive vibrations, overheating, and so on). The operator should look for obvious wear and tear, broken and damaged parts or cables, damaged ductwork, paint peeling off, and so on. In case where serious damage or leaks are found, countermeasures must be taken by a skilled-operator or technician. All abnormalities or defects found should be recorded and remedial action taken.

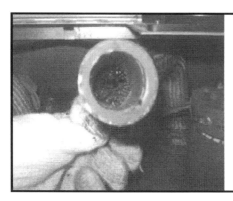

Figure 6-4: A dirty exhaust duct which was not cleaned for years.
This only came to light when dirt and debris started to show up on finished parts that were heated in the oven.

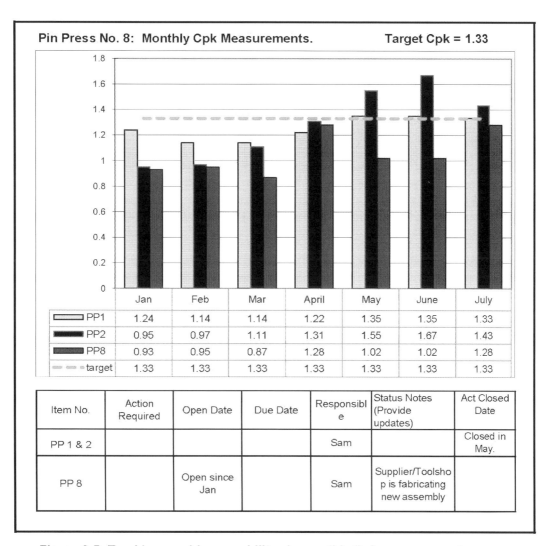

Figure 6-5: Tracking machine capability via monthly Cpk measurements

6. **Determine preventive maintenance standards, document, and distribute to TPM team.** Determine complete maintenance standards for all machines. This includes recommendations from the machine manufacturer. Include cleaning, lubricating, inspection, and part replacement standards.

    Use manufacturer's maintenance schedule and parts replacement guidelines. List what is to be done by operators (daily or weekly) and by technicians (per a schedule). Keep a complete record of maintenance performed, deterioration of specific machines, and remedies. Another important point is to understand where the equipment is in its life cycle as this will influence and determine the inspection and maintenance schedule; we need to refer to the manufacturer for guidance in this area.

7. **Use diagnostics to forecast potential failures.** This is where we need skilled and trained technicians to use software or documented diagnostics provided by manufacturers to review or forecast part replacements. Diagnostics are available for most complex machines. This includes diagnostics and software to check for machine capability (Cpk) and equipment accuracy. Such diagnostics should be run per the manufacturer's schedule.

8. **Implement product quality assurance and machine process capability.** Monitoring and measuring the quality of the finished product is crucial; for example if we see a deterioration of product quality, it would imply that the machine is producing defects; consequently we must go to the machine and look for the source of the degradation. Alternatively, we can look at machine process capability, to understand if the machine or process is under control and capable of delivering the expected quality per the specifications.

    a. **Product quality assurance:** To get started we must determine the criteria for monitoring product quality assurance. Specifically:

    i. The quality parameter of the finished part must be quantitative and clear and any variance should be quantifiable. For example a dimension, surface finish, or a characteristic easily identified and measured. Parts failing test will be scrapped – which is a standard measure of machine quality.

    ii. Use SPC (statistical process control) run charts to monitor product quality. For example, we could use a SPC run chart to monitor the dimensions of a finished part. As long as this stays within defined limits, we can conclude the machine is running smoothly. If we begin to detect slow degradation, we need to check for the root cause and provide a remedy.

    b. **Machine process capability:** Another important criterion is machine process capability: This measures the ability of the machine to produce parts within defined limits. However, the concept of process capability only holds if the process or machine is running in control for a period of time. When we measure machine process capability we are able to predict if the machine will produce good parts. Specifically:

   i. Process capability can be measured by monitoring machine accuracy, tolerance, thermal profile parameters, and so on. We can track the process capability, or Cpk, of key machines. Manufacturing specifications call for annual checks for Cpk on many complex machines. In the interim monitoring the product quality characteristics may be sufficient and less expensive. In some critical machines, the Cpk has to be monitored and measured regularly.

   ii. Refer to Fig. 6-5 where we show a chart from an operation running a pin-press. Here the operation started tracking Cpk of the press as an alternative to measuring finished parts; when the Cpk was found to be too low, countermeasures were taken to bring it to the target of Cpk = 1.33. Once the process capability was above target, it was concluded that the machine was running smoothly; this was verified by the quality of the finished product.

   c. Both product quality assurance and machine process capability are good quality assurance measures. However, in the spirit of a focused effort and cost control, we recommend checking Cpk on a periodic basis per manufacturer's requirements; the exception would be when you have a sensitive machine that needs routine checks, as in the example here.

9. **Reduce setup and changeover times to ensure maximum machine productivity.** We discussed this in an earlier chapter. Quick setups and changeover will increase machine availability, reduce production lead time, and improve the OEE factor of the machines.

10. **Monitor and measure overall equipment effectiveness (OEE).** For measuring equipment efficiency we recommend using the OEE measure which we discussed earlier. Check that the machine is within its design specifications during use and is maintained per manufacturer's guidelines. If not then maintenance, calibration, or technician skill-sets may be an issue.

    a. To keep track of individual machine performance, OEE is a good measure of machine effectiveness. World class performance is for OEE to be in the range of 85-95%. When you are working to improve machine efficiency using the OEE metric, it is best to start by looking at bottlenecks created by machines: these should be fixed one at a time, moving from one bottleneck to the next.

    b. Since OEE has three parameters the impact from any substandard parameter is the same to production output or the customer: Production targets or shipments will be missed. However, internally, a substandard Q parameter will be the most expensive due to the rework required. Hence, it's best to first improve the Q parameter first and target for higher than 95%. Then work on the A parameter or availability: Review machine availability, which depends on machine uptime, good maintenance, and operator availability to run the machines.

c. Finally look at the P parameter or machine performance: Is it delivering per manufacturer's specifications? This is often termed nameplate or cam speed. Sometimes, technicians will run a machine below nameplate speed because of machine issues; this could be a machine whose process capability (Cpk) has deteriorated due to poor maintenance or a machine that is being asked to deliver specifications beyond its capability.

d. In Fig. 6-6, we show the OEE calculation for a press. This was part of an exercise for a new production setup and run. The OEE was calculated to ensure that the machine would deliver parts per the customer's daily production requirement. The production run was planned for one shift with no overtime hours. As we discussed earlier, the quality parameter must be high to ensure high OEE; in this case it is almost 99%. Note that the performance efficiency is running higher than 100%. How is that possible? It was higher because the theoretical cycle time was no longer accurate – this could be due to several reasons including the fact that the machine was recently upgraded per the manufacturer's advice. In this case the theoretical performance needs to be revised.

e. In Fig. 6-7, we show a chart monitoring OEE for two adjacent machines. As can be seen it was a challenge to maintain the target of 85%. Because there were three parameters to control within OEE; it required a strong effort, good TPM plan, and high product quality. According to the engineer running these machines, he mentioned that product quality was running at a high level of 98%; however they had issues with machine and parts availability which needed to be resolved.

TPM for machines is a requirement to ensure smooth, non-stop, high quality production. TPM in aircraft, however, is a much bigger challenge: Besides running smoothly and efficiently they also have to ensure no lives are lost. The attached box illustrates some of the TPM strategies for aircraft.

Overall Equipment Effectiveness: Machine M1
(Calculated after 24/08 Production Run)

Description	Formula	Data	Target
Equipment Availability			
Total Scheduled Time (mins.) (A)	A	480	
Operator breaks/lunches causing downtime	B	15	
Net Available Time (mins.) (A-B)	C = A - B	465	
Planned downtime, breakdown or maintenance	D	39	
Operating Time (mins.) (C-D)	E = C - D	426	
Availability (E/C)	F = E / C	92%	95%
Equipment Performance Efficiency			
Total Parts Run	G	461	
Theoretical Cycle Time (mins/part)	H	0.95	
Performance Efficiency ((G x H) / E)	I = (G x H) / E	103%	95%
Quality of Parts			
Total Rejected Parts	J	6	
Quality ((G - J) / G)	K = (G - J) / G	99%	99%
Overall Equipment Effectiveness (OEE) = (F x I x K)		93%	89%

Fig. 6-6: Measuring OEE performance

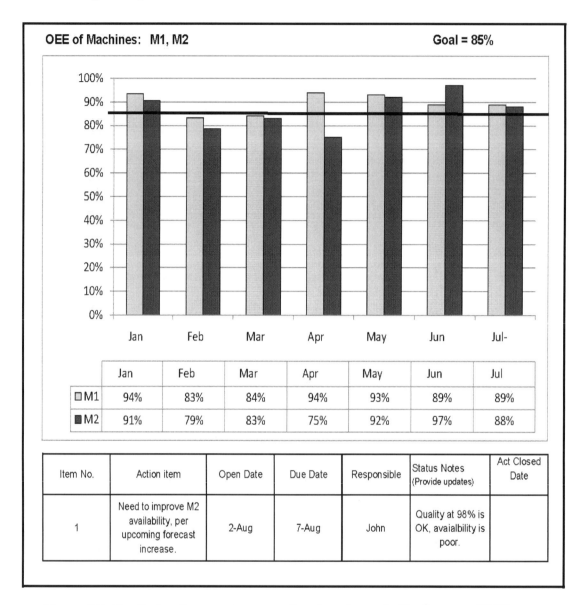

Figure 6-7: Monitoring & managing operational equipment effectiveness (OEE)

Summary: Machine Management: Quick Setup & TPM

We expect machines to provide predictable performance. To this end we have discussed several areas of machine management, including how to reduce production lead times via quicker setup procedures and TPM (total productive maintenance) methodologies.

Production lead time for manufacturing processes can be reduced by reducing machine setup time. This is the purpose and impact of the of the SMED program, which we discussed in detail. Shorter setup times means that machines can minimize the production lot size and reduce production inventories. This will allow for more efficient just-in-time manufacturing; this will also enable an operation to be efficient in high-mix low-volume production. Overall it will be a more responsive and cost-effective operation and be able to respond quickly to small customer orders. This can open an operation to new business and market segments.

We must also ensure that machines work every time. For that we need a TPM plan. The objective of TPM is to ensure that machines provide predictable performance, deliver exactly what we expect of them, with zero (or low) errors, breakdowns, or accidents; all this at a reasonable cost.

Machines must also build high quality parts: In the next chapter we discuss how to build quality parts and processes

Chapter 7

Quality: Improvement and Control

*Create constancy of purpose for continual
improvement of products and service to society.*
Edwards Deming

Overview

Product and service quality is essential for any business. To this end we need quality tools and systems to manage the factory floor, support services, and the overall business. The benefits of this effort will help ensure efficient processes, higher productivity, and less fire-fighting. This will result in more time for innovation and creativity, improved quality of products and services, lower costs, and increased customer satisfaction and loyalty.

In this chapter we will review how to drive quality improvements in a systematic and analytical way, discuss methods to control and manage quality of existing products and processes, and review methodologies for aiming for zero defects. Specifically, we will cover:

- Quality and customer satisfaction.
- Continuous improvement and the PDCA cycle.
- Kaizen, or continuous improvement, activity.
- Management PDCA: Using the A3/A4 process.
- Problem solving hierarchy.
- Quality control and management.
- Statistical process control (SPC).
- Monitoring and managing quality on the production line.
- Aiming for zero defects: Recommended methodologies.
- Setting quality goals.
- Total quality management (TQM).

Quality and Customer Satisfaction

Quality Definition

Before we start a quick review of Quality and customer expectations is appropriate. There are numerous definitions of quality. In our context here is the one we prefer: *Products and services that meet or exceed customers' expectations.* The key words here are exceeding customer expectations, which can be very demanding and will be relative to what other manufacturers or companies provide.

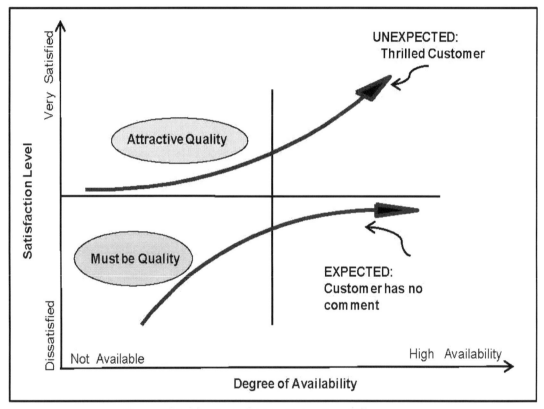

Figure 7-1: The two dimensions of quality

Dr. Noriaki Kano has proposed a two dimensional model of quality[22]: *"Must be quality"*, or a set of expected features, such as reliability, and "*Attractive quality*", or the unexpected that goes beyond the customer's needs. *Attractive quality* consists of the extra features or value that the customer would love to have but has not yet desired because he or she has not yet thought about them. Refer to the discussion in the accompanying box and Fig. 7-1.

Customer Satisfaction

The Kano model gives us a clue on what we need to do to ensure high customer satisfaction. We need both reactive and proactive activities. The reactive approach is necessary to understand and resolve challenges and problems arising from current products and services. The pro-active approach is essential to help influence and create new products and services. We briefly touch on both approaches here.

The Kano Model: Two Dimensions of Quality

Noriaki Kano and others have proposed the concept of two dimensions of quality: *"Must Be Quality"* and *"Attractive Quality"*. Refer to Fig. 7-1: This is often called the Kano Model. In the figure, the vertical or Y axis represents the customer satisfaction level or performance - the higher the better. The horizontal or x axis represent the degree of availability of the characteristic, in this case *must be* or *attractive*.

"Must Be Quality" is that aspect of a product or service which the customer expects. If the customer does not get it, she will be extremely dissatisfied. However, if the customer gets this feature, she will have no comment and will be moderately satisfied. Examples of this are a reliable, safe, and easy to use product. This is the minimum acceptable standard.

"Attractive Quality" is that aspect of a product or service that goes beyond current needs. In Japanese, this attractive quality feature is called *miryokuteki hinshitsu*. This is the quality or value that the customer is not expecting, but receives as a bonus. The customer will be thrilled and excited, but if the feature is not available, the customer has no comment. However, he may purchase an alternative product with additional features or value. An example of this could be a preference for a car with a sun roof or an Apple I-Phone Vs another mobile phone. With time, such *"attractive quality"* becomes *"must be quality"*. An example is the anti-lock brakes, which today are an essential or "must be" quality item in most cars. Similarly when staying at a hotel, we get acquainted with "must be" and "attractive" or value items. Such differences distinguish a good hotel from a mediocre one, given that price is comparable.

A well designed product or service should have both dimensions of quality as these can strongly influence the customer's buying decision and satisfaction. A customer satisfaction model can be created from these dimensions. The first step in creating this model is to understand the quality characteristics that the customer wants. These quality characteristics are termed the Voice of the Customer. Once the Voice of the Customer is understood, we can attempt to translate it into quantitative terms in our product or service offering.

Reactive activities

A system to manage and resolve customer issues and complaints. According to Theodore Levitt: "One of the strongest signs of a bad or declining relationship is the absence of complaints from the customer. Nobody is ever that satisfied, especially not over an extended period of time. The customer is either not being candid or not being contacted."

Therefore it is important we have system to record and resolve customer complaints quickly as they occur; more important, we should have a method that discovers root cause of the issue and prevents recurrence. Customer inputs can also be collected via routine surveys, such as a post-purchase survey, a post-installation survey, customer satisfaction survey, and customer inputs via face-to-face meetings.

Competitive benchmarking. Our purpose here is to look at manufacturing processes and technology used by other leading companies and to use that learning to improve our processes. At the same time, we cannot just copy others. Therefore we should look for new technology at exhibitions, seminars, and university research departments, with the purpose of detecting trends, which allow us to understand or predict the next successful technology, product, or service.

Proactive activities

Developing and implementing "Attractive Quality" or the "Total Product or Service" concept. We have discussed the concept of attractive quality. Harvard Business School Professor Ted Levitt provides a handy idea, which he labels the "Total Product Concept."[23]

The total product or service concept is as follows: Picture four concentric circles: The inner one is labeled "*generic product*", next comes "*expected product*", then "*augmented product*." The last, which has no boundary, is labeled "*potential product*. This concept applies to every product or service.

For a manufacturing environment that wants to be World-Class, the "*generic product*" is what every manufacturer is able to achieve: Build per specifications and agreed upon cost; all operations must be able to do this.

But the customer appreciates additional services, hence for "*expected product*" it can include: Excellent safety record, excellent work environment for employees and customer, ability to prevent potential quality and manufacturing issues, working with and managing processes per product safety and ISO requirements, continuous cost reduction, and improving customer satisfaction.

For "*augmented product*" it can include: Ability to forecast and manage supply and demographic issues, manage attrition with no impact to product or process, and direct delivery to end-customers. It would also include a review of the products it manufactures with improvement in design and manufacturability. This requires a full review of the product to improve productivity, improve quality, and cost reduction to help the design team.

For *"potential product"* it can be: Total product and process management, with no further R&D or OEM customer involvement required, complete handling of reverse logistics until product obsolesce, and design services.

This is what the customer wants: A total Product, not just basic manufacturing capability.

Quality Improvement and the PDCA Cycle

In this section we discuss continuous improvement and the PDCA improvement cycle; specifically we look at the steps of the cycle, discuss it in great detail, understand how to do it well, and review its various applications. Several applications of the PDCA cycle have come out lately, for example A3 Reporting, which has generated texts and two day seminars. However, if we understand the PDCA cycle, we can comprehend and put into practice all of these applications.

Problem Solving Styles

How many ways can you solve a problem? Based on our experience, we show in Fig 7-2, a cluster of styles that have evolved over the years. In all three styles we show the reaction to a problem, how the root cause is determined, and the solution implemented. From this list, we obviously recommend the analytical approach, which uses the PDCA improvement cycle.

The PDCA Cycle

At a seminar we once attended, a participant asked the legendary Dr. Kaoru Ishikawa, considered the father of Japanese Quality:

"What is the most important tenet of Quality?"
He replied without hesitation: *"The PDCA cycle."*

This is an important point: The PDCA cycle is a systematic, data driven, analytical, and scientific process which is used to improve or manage products, services, or processes. The PDCA Cycle is shown as a circle or wheel – this implies that you keep going around in a process of *continuous improvement*. Refer to Fig. 7-3.

The PDCA (Plan Do Check Act) cycle was originally developed by Walter Shewhart – the originator of Statistical Process Control. It was popularized by Dr. Edwards Deming[24] and is often called the Deming Cycle. The Plan Do Check Act cycle is stated as follows:

Plan: Determine goals and methods to reach the goals.
Do: Educate employees and implement the change.
Check: Check the effects. Have the goals been achieved? If not, return to the Plan stage.
Act: Take appropriate action to institutionalize the change.

The Shotgun Approach:

Method: The cause of a problem is determined from gut feel, intuition, experience, or intelligence. The solution, to eliminate the cause, is then implemented. This method is also called the genius approach.

What are the pros and cons of this method? The solution can be implemented quickly. The effectiveness of the solution depends on an individual's or team's intelligence or experience level. We have observed brilliant individuals, who have come up with very effective solutions. In most cases, however, the solution is often wrong because it misses the root cause.

Who uses this method? Many people. Typically, this method is used by many politicians, some senior managers, and non-analytical people. Definitely not recommended, except for very experienced or brilliant individuals.

The Explorative Approach

Method: The problem is explored in detail and the individual or team goes through a winding, tortuous, journey through many ideas, data, and alternate causes and solutions. One of the solutions is then selected from the alternatives, fine-tuned, and implemented.

What are the pros and cons of this method? There are many good aspects in this method, including the gaining of group consensus and generation of creative solutions. This method takes medium to long time but there is no systematic approach on how to proceed, hence, the wrong solution could be implemented.

Who uses this method? This method is used by intellectuals and creative people. Why? We suspect the reason is that they favor a free-flow of ideas in order to get innovative solutions. However, such an approach can range from great, to complex, to suboptimal.

The Analytical Approach

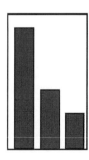

Method: The individual or team goes through an analysis and background of the current situation. Then they search for and validate the root causes of the problem. Finally solutions are identified and implemented.

What are the pros and cons of this method? This method generates a powerful solution that gets to root cause of the problem, hence there is little possibility of a recurrence. This method can be taught and documented. It takes medium to a long time but many people find it frustrating to proceed in such a systematic fashion, which requires discipline to stay on track.

Who uses this method? It is used by analytical people and those trained in quality improvement techniques. Many Japanese managers have been observed to use it, probably because of Japan's history and culture of aggressive quality improvement. It is also popular in quality oriented companies and organizations.

Figure 7-2: Problem Solving Styles

Detailed PDCA cycle

Next we discuss the detailed PDCA cycle. This discussion is useful as a reference for training. Each of the seven steps in Fig. 7-3 is explained in greater detail. In addition, we have listed some quality control (QC) tools that could be used in each step. Following the discussion we discuss a completed project.

PLAN Stage

Step 1: Select the theme or project

Objective of this step: To clearly define the problem to be resolved.
Discussion: Here we define the project; understand its background, set a target, and prepare a schedule of activities. This is the most important stage and hence every step needs to be done thoroughly. The following are sub-steps of step 1.

Step 1a: Project background and reasons for selection.

- The project background must be very clear: the project can be selected from the department objectives, customer complaints, or area selected for improvement.
- At the start of the project, the project team must be formed. The team should be cross-functional and include the relevant experts in process, production, and quality. A team leader must be appointed to drive the team in going through all the key steps of the PDCA and to ensure the team works diligently towards meeting the schedule and target. The team also needs a record keeper and, if necessary, a facilitator or champion.

Step 1b: Set the objective and target of the project.

- This should include a statement of the project objective, a numerical target to be achieved, and a time frame for project completion. For example: Improve product test yield to 99 %, by May 2012.
- Always set reasonable and realistic targets. The target can also be set after step 3, when more data is available. The situation will vary with each project, but we recommend that a tentative target should be set here.
- Set the target of the project with these guidelines:
    - Use current data to set an aggressive breakthrough goal.
    - Use competitive data to equal or better the competition.
    - Use the rule of thumb of reducing defects by 50 percent every six months. Refer to the sidebar later in this chapter on setting targets.

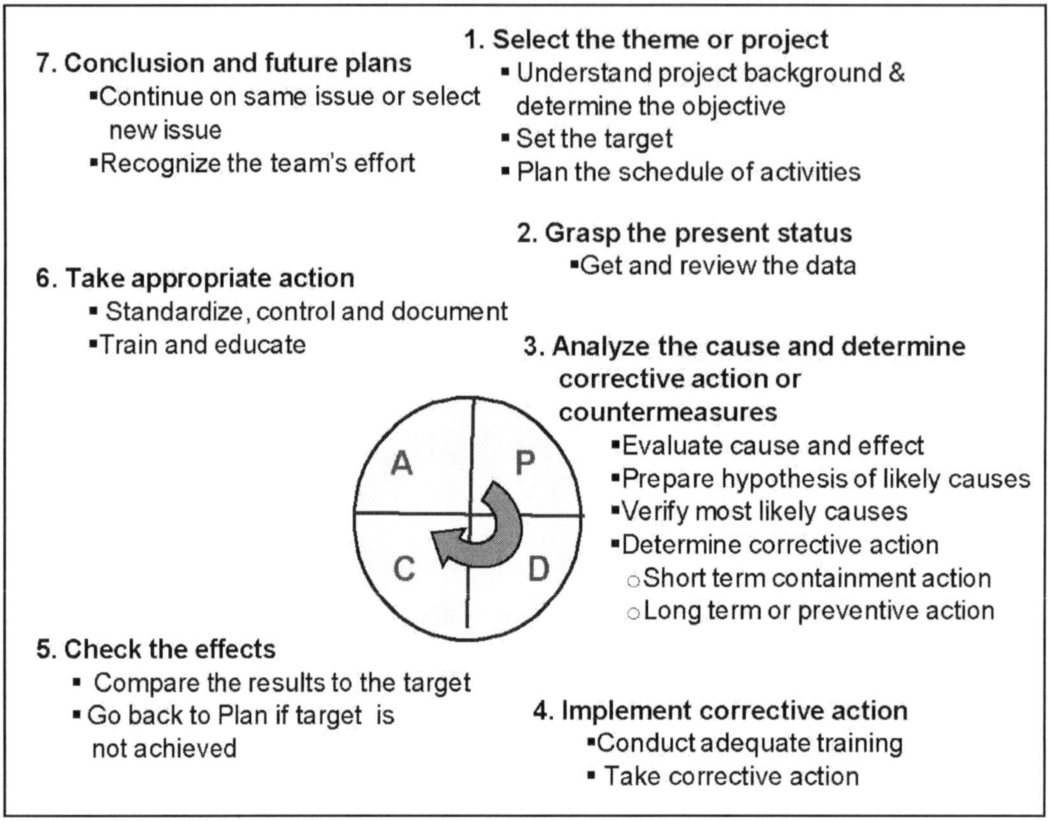

```
7. Conclusion and future plans          1. Select the theme or project
   ▪Continue on same issue or select       ▪ Understand project background &
    new issue                                determine the objective
   ▪Recognize the team's effort          ▪ Set the target
                                         ▪ Plan the schedule of activities

                                         2. Grasp the present status
                                            ▪Get and review the data
6. Take appropriate action
   ▪ Standardize, control and document
   ▪Train and educate                    3. Analyze the cause and determine
                                            corrective action or
                                            countermeasures
                                            ▪Evaluate cause and effect
                                            ▪Prepare hypothesis of likely causes
                                            ▪Verify most likely causes
                                            ▪Determine corrective action
                                              ○Short term containment action
                                              ○Long term or preventive action
5. Check the effects
   ▪ Compare the results to the target
   ▪ Go back to Plan if target is        4. Implement corrective action
     not achieved                           ▪Conduct adequate training
                                         ▪ Take corrective action
```

Figure 7-3: The PDCA Cycle

Step 1c: Prepare a schedule of activities.

This lists the steps in the PDCA cycle and the expected time frame for each step. A good rule of thumb is project completion in less than six months. With experience the team can gauge the project complexity and the required timeframe to complete. It is important that that this schedule is taken seriously.

QC tools that can be useful for step 1: Trend Chart, Pareto diagram

Step 2: Grasp the current status: Get and review the data.

Objective of this step: To understand the problem and to highlight specific problems.
Discussion: Here we study the effects of the problem, by reviewing the available data. Our study should be approached from several facets, *such as, time, location, and type*. For example if we want to reduce the percentage failures during the production of a fax machine:

- For time, we can look at the failures between the day and night shift, and over a period of time.

- For location, we can look at which part of the machine has most failures – for example top, bottom, side center, pc assembly, or mechanical parts. The available data can be presented in graphs and Pareto charts.
- We should also get a *process flow chart* of the product or process being studied. If it does not exist, we must prepare a chart. This will help understand the current process and highlight where in the process problems occur; it will also draw attention to inventory, flow, environmental, and test issues. After the process is improved, a new revised process flow chart will have to be drawn.

Additional notes for Step 2:

Grasp the current status using the 5W/2H or 5W method:

The 5W/2H method: This procedure will help drill down into the problem.
Discussion: Here we study the effects of the problem using the 5W and 2H method, for example:

- **Who:** Identify individuals, processes, and products which have the problem
- **What:** Describe the problem, get the data, and plot a Pareto diagram.
- **Where:** Identify where on a product or location the problem occurs, use photos, checksheet, or diagrams. Also check if it is geography or demographic related.
- **When:** Identify the time this occurs, e.g. during first or second shift operation, time of month or year; could it be seasonal problem?
- **Why:** Do we know why it is happening?
- **How:** Do we know what procedure was used to cause this to happen?
- **How:** How large is the problem? Do we know if the process is in or out of control?

The 5W method: We can also use the 5W method; here the approach is to ask Why 5 times in order to drill down deep to understand how the problem occurred. Let's look at an example of the 5W analysis made for a failed motorized window of a car:

- Why was the car window not working?
- We opened the door panel and found a damaged cable harness wire.
- Why was the cable harness damaged?
- It was cut and damaged because it was trapped between the door panel and the door body when the panel was screwed in.
- Why was it trapped by the panel, hence damaged?
- The harness was very near to the edge of the door panel cover (refer to pictures) and protruding out of the door well.

- Why was the wire protruding out of the door well?
- Currently the operator has to judge if the cable harness is in the correct position before screwing in the door panel.
- Why does the operator make a judgment call?
- The standard work instruction does not give any guideline; it only provides a quality alert or warning: *"be careful that the harness does not protrude out before you screw in the panel"*.
- OK, let's review the standard work document……..

The 5W method allows us to home into the problem or it gets you started in the right direction. It is useful for less complex problems or as a high level approach by managers. Nevertheless, there is no substitute for good data collection and analysis.

Important points for Step 2: Grasp the current status: According to Spear and Bowen[25], in reviewing production errors at Toyota Motors, typical failure modes or causes for current problems can often be traced to failures in 3 specific areas:

- Standard work (work sequence, work timing, or defective material).
- Customer-supplier connection, communication, and capability are clear regarding mix, demand, specifications, and ability to manufacture.
- Clear, simple, and direct pathways for every product and service resulting in good workflow.

QC tools that can be useful for Step 2: Process flow charts, Pareto charts, trend charts, control charts, histograms, process capability indices.

Step 3: Analyze the cause and determine corrective action

Objective of this step: To determine the root cause of the problem, and plan for corrective action.

Discussion: In this step we examine the causes of the problem, isolate the root causes, and determine corrective action. The detailed sub-steps (3a-c) are:

Step 3a: Prepare cause and effect diagram. The item to examine is selected. This can be the first or second bar in a Pareto diagram of defects, or it can be the first three bars. Other times we may select a specific item we want to improve. A cause and effect diagram is then prepared.

- The causes in the cause and effect diagram can be derived through a brainstorming session. The brainstorming session should be free-flowing and all possible and impossible causes should be listed.
    - A C/E (cause and effect) diagram can be used for any type of problem.
    - A detailed and complex C/E diagram indicates a good understanding of your process and technology.

- o   A simple C/E with a few causes (say 4 or 5) may indicate poor understanding of your process or technology.
  - o   Hence, grouping possible causes into several categories, such as Man, Machine, Method, and Material (for a manufacturing environment) is recommended to facilitate drawing up possible causes.
  - o   Other environments may require different categories.
- Next, use data that was obtained in step 2, to eliminate unlikely causes. The cause and effect diagram can now be simplified and re-drawn, and the unlikely causes crossed out.
- Depending on the problem encountered, other tools that can be used include a FMEA analysis, matrix diagrams, and 'five whys'.

Step 3b: Prepare a hypothesis and verify most likely cause

Discussion: We prepare a hypothesis by selecting the most likely causes out of the cause and effect diagram; this can be done using the group's experience or voting.

- This list of most likely causes must now be verified with data.
- But we should use new data to determine if there is a relationship between the selected causes and the effect; this may require us to conduct experiments.
- We will now have a short list of verified causes – the root causes of the problem we want to reduce or eliminate.

Step 3c: Determine corrective action.

Discussion: We must now decide on the corrective action. Sometimes the corrective action is obvious. If not, we need to decide on the action. Creative alternatives should be generated using brainstorming or cause and effect diagrams. There are 2 types of corrective action or countermeasures:

1. **A short term fix or containment action:** This could include immediate 100% inspection for the defect and rejection of defective units, or inspect for and repair of the defect. This is typically required at the beginning of a critical failure or for high customer complaints; but may be skipped during a routine continuous improvement project. **Note:** This is not the solution, only a containment action because we still have to find a more effective solution.
2. **A long term fix or preventive action:** This will include elimination of root cause, hence preventing the problem from recurring. This is essential, but due to constraints of implementation the quick fix may be implemented first. It will be necessary to conduct a trial of the proposed action, to determine that it works. Only then should it be implemented. For the preventive action for complex projects, we recommend the use of an implementation plan.

QC tools that can be useful for step 3: Check-sheet, checklist, implementation plan, stratification, and statistical design of experiments, Pareto diagram, cause and effect diagram, FMEA analysis.

DO Stage

Step 4: Implement corrective action.

Objective of this step: Implement the plan and eliminate the root causes of the problem.
Discussion: Employees who execute the correction must understand the corrective action. Good communication and training will be necessary. The following are recommended sub-steps:

Step 4a: Prepare instructions and flow charts for complicated procedures.
Step 4b: Adequate training must be provided.
Step 4c: Follow the plan exactly.
Step 4d: Record any deviations from plan and collect data on results.
QC tools that can be useful for step 4: Checklist, check-sheet, trend charts, standard work.

CHECK Stage

Step 5: Check the effect of corrective action

Objective of this step: To check the effectiveness of the corrective action.
Discussion: We now check the effect of the corrective action. There are several sub-steps that must be followed:

Step 5a: Compare overall result.

- Here we review the overall results. We can review improvements on a paired Pareto diagram in order to compare before and after performance.
- Before and after results should be compared on all other items selected for study in Step 2; use the same tools for making the comparison, such as bar graphs, paired Pareto, trend chart, control charts, histograms, and process capability indices.

Step 5b: Failure to meet results.

If failure is due to improper implementation, then we must go back to Step 4, implementation. Otherwise, we go back to Step 3, analysis. If we fail to meet our goals, it is very likely that we missed the root causes, and further analysis will be required.

Step 5c: Results have been achieved, goal has been met.

If the overall result is equal or better than the target set in Step 1, we review the before and after data – especially the Pareto diagrams – and check that there are no side effects, that is, there is no increase in the other categories of failures.

QC tools that can be useful for step 5: Paired Pareto diagram, trend charts, control charts, histograms, and process capability indices.

ACT Stage

Step 6: Take appropriate action

Objective of this step: To ensure that the improved level of performance is maintained.
Discussion: The corrective action that has been successful in improving performance must be documented in current operating procedures. There are several sub-steps:

Step 6a: Documentation, standardize and control.

- The corrective action (implemented in step 4) that has been successful in improving the performance level should be documented in current operating procedures or standard work. Poor documentation can result in problem recurrence in the future. It is very important to convey this information to other parts of the organization, which may have generated the root cause of this problem.
- It is also important to identify critical process parameters to control. Hence the standard work instructions, production control plan, or quality checkpoint chart should be updated.

Step 6b: Training. Ensure appropriate training in the new standard work. Employees must fully understand the changes that have taken place and the new procedures.

QC tools that can be useful for step 6: Trend chart, control chart, check-sheet, standard work instructions, control plan/quality checkpoint chart.

Step 7: Decide on future plans.

Objective of this step: Use the experience gained for future projects.
Discussion: An area where the next project can be found is the results in step 5.

- If our new trend chart or Pareto diagram has sharp peaks, we must eliminate them. If the bars in our new Pareto diagram have even heights, we must change our stratification base, and then decide what to eliminate.
- If we have created side effects, then we must work on eliminating them. We may also start afresh with a new breakthrough activity.
- But, first, we make sure that the entire process we have been through is documented according to the seven steps listed here.
- The decision to continue on the current project or to select a new one has to be based on priorities and resources.

116   Winning with Operational Excellence

Project: Improvements Using PDCA Methodology

We show and discuss here a completed project, which comes from Hewlett-Packard. This product is a small display panel, used in cars and computers. The team that worked on this project was very thorough and systematic, and won an internal award from management for achieving excellent results.

Step 1: Select the Theme or Project

Step 1a: Understand the Project background and determine the project objective.

To improve first pass yield at final test for product QDSP-678Current status and issues are:

- Low first pass yield at final test (86.2%).
- Too much rework at final test.
- Shipment timeliness adversely affected because of low yields and rework.

Step 1b: Set a target.

To achieve 96% first pass yield at final test by July 2xxx. This is equivalent to reducing failures from 13.8 % to 4 %, and in line with the department goal

Step 1c: Prepare a schedule of activities.

The project schedule is shown in Fig. 7-4. Note that the timelines follow the seven steps of the PDCA cycle.

Activity	wk wk wk wk wk wk wk wk wk wk wk wk wk wk wk wk wk wk wk
1. Select The Project or Theme	
-Set objective and target	
-Schedule activities	
2. Grasp the present status	
3. Analyse cause, determine corrective action	
-Evaluate cause & effect	
-Prepare hypothesis	
-Verify most likely casues	
-Determine corrective action	
-Short term, containment action	
-Long term, preventive action	
4. Implement corrective action	
-Take corrective action	
-Conduct training	
5. Check the effects	
-compare the results to target	
6. Take appropriate action	
-Standardize, control, document	
-Train and educate	
7. Conclusion and future plans	

Figure 7-4: Project Schedule

Step 2: Grasp current status

In Fig. 7-5 we show a graph and Pareto diagram. The graph gives the final test yield data for the display assembly, and the Pareto diagram gives a breakdown of the causes of failures. A process flow chart is available but not shown here.

For the Pareto diagram categories: Open Digits means any digit that does not light up. Wrong sequence means the displays fail to exhibit the correct sequence per product specifications.

Note: The *other* category of a Pareto diagram is usually on the right hand side of the diagram. Often, when the other category is large, it shows up on the left hand side or in between other categories – this is not correct, it's best to break down the *other* category if it is large.

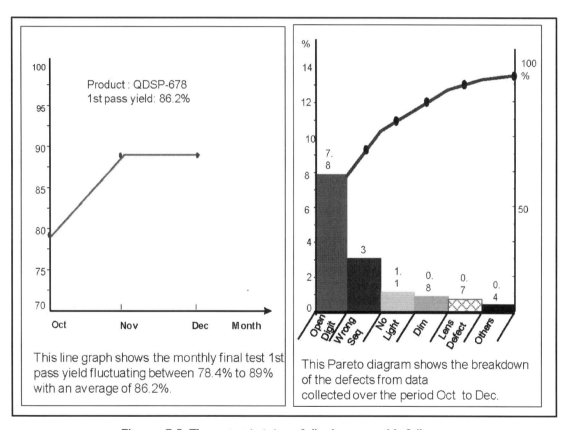

Figures 7-5: The current status of display assembly failures

> **Step 3: Analyze Cause and Determine Corrective Action**
>
> **Step 3a: Prepare cause and effect diagram**
>
> The highest fail categories in the Pareto diagram – open digits and wrong sequence – were identified and a cause and effect diagram prepared for each item. The team brainstormed for possible causes of each item, using the 5W/2H method. The cause and effect diagrams are shown in Fig. 7-6 and 7-7.

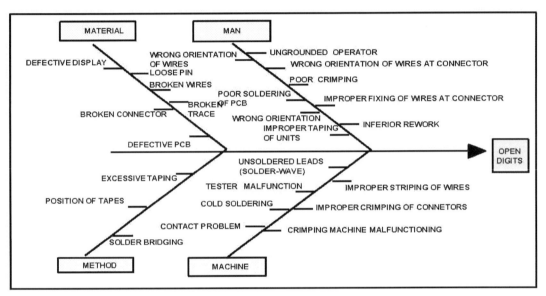

Figure 7-6: Cause and effect diagrams for open digits

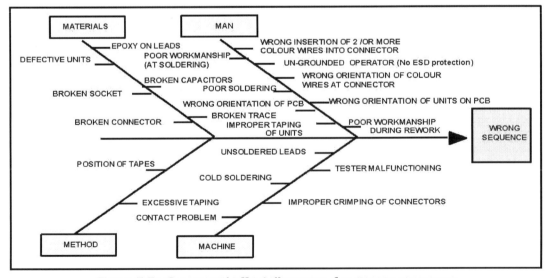

Figure 7-7: Cause and effect diagrams for wrong sequence

Step 3b: Prepare a Hypothesis and Verify Most Likely Cause

The team deliberated on the possible causes and then voted for *the most probable causes,* based on their job experience and knowledge. They were:
1. Open Digits
    a) Wrong orientation of wires inserted into connector.
    b) Improper fixing of wires into connector.
    c) Broken wires.
    d) Unsoldered leads.
    e) Defective units.
2. Wrong Sequence
    a) Wrong orientation of wires inserted into connector.
    b) Wrong orientation of wires at soldering.
    c) Improper crimping of connectors.

Verification: Additional failure data was collected in order to verify which of the possible causes were most likely. The analysis findings were summarized in Fig. 7-8. Members noticed that most of the causes were workmanship related. The verified probable causes were:
1. Wrong orientation of wires inserted into connector.
2. Wrong orientation of wires at soldering.
3. Improper fixing of wires to connector.
4. Unsoldered leads.

Type Of Defect	Open Digit	Wrong Sequence	Open Digit & Wrong Sequence	Total
Wrong orientation of wires at connector	16 (53.3%)	13 (76.5%)	4 (100%)	33 (64.7%)
Wrong orientation of wires at soldering	4 (13.5%)	4 (23.5%)		8 (15.7%)
Improper fixing of wires to connector (wires get loose)	3 (10%)			3 (5.9%)
Unsoldered leads	4 (13.3%)			4 (7.8%)
Broken wires	1 (3.3%)			1 (1.96%)
Defective unit	1 (3.3%)			1 (1.96%)
Improper crimping of connector	1 (3.3%)			1 (1.96%)
Total Quantity Analyzed	30 Sets	17 Sets	4 Sets	51 Sets

Figure 7-8: Verification of most likely causes

Step 3c: Determine Corrective Action

Short term fix, or containment action: Since this is an improvement project and not a response to a customer complaint, there is no containment action. The long term fix or preventive action: Since the four major causes were workmanship related, the team attempted to list down the snags and difficulties in the processes that are related to each of the causes. Members then proposed and discussed what could be good solutions and preventive actions for each of these. This is summarized in Fig. 7-9. Members also discussed the requirements for each of the work holders and soldering block, and had a volunteer draft out the preliminary drawings. A final meeting was held with the vendor, after which the vendor finalized the drawing, selected the material, and fabricated the respective fixtures.

	Causes		Snags & Difficulties		Recommended Action
1	Wrong orientation of wires inserted into connector	a	Insertion of the 16 wires in their correct sequence into the connector wad done with the aid of the reference chart and by memory		Design a work-holder for the connector with the appropriate color of the 16 wires painted on it to act as guide for correct insertion of the wires.
2	Wrong orientation of wires at soldering onto PCs.	a	Plastic bags were used to store the 16 colors	a	Design on appropriate work-holder at soldering for the storage of the 16 wire types that arranges them in the appropriate sequence.
		b	Soldering of the 16 wires in their correct sequence into the PCB was done with the aid of a reference chart and memory	b	Modify the soldering block to include color guides for the 16 wires.
3	Improper fixing of wires into the connector (wires came out)	a	Each crimped wire should be fully inserted into the connector until there is 'click' sound.	a	Implement a "pull test" at wire insertion for every wire to ensure that each crimp is correctly seated in the respective sockets.
4	Unsoldered leads.	a	There is limited space on the PCB for tape adhesion when taping the display components for wave solder.	a	Implement 100% inspection on all taped units before wave soldering to ensure that all the corner leads are not covered with tapes so that all leads will be soldered during wave soldering.

Figure 7-9 Determine corrective action

Step 4: Implement Corrective Action

A training session was held for all production staff to educate them on how to use the fabricated work-holder and the soldering block. The recommended action was then implemented in and results were tracked every day. The results were tracked on the graph shown in Fig. 7-10.

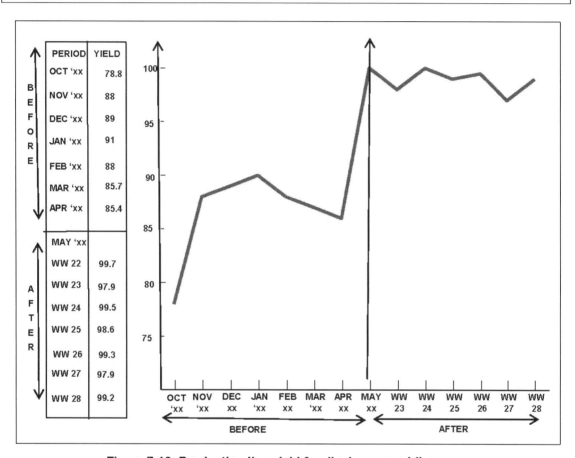

Figure 7-10: Production line yield for display assemblies

Step 5: Check Effects of Corrective Action

A detailed analysis was done. The data was then plotted on a paired Pareto diagram. The results show a marked improvement in open digits. A check was made for unwanted side effects. There were none – in fact some of the other Pareto bars also decreased, refer to Fig. 7-11.

122 Winning with Operational Excellence

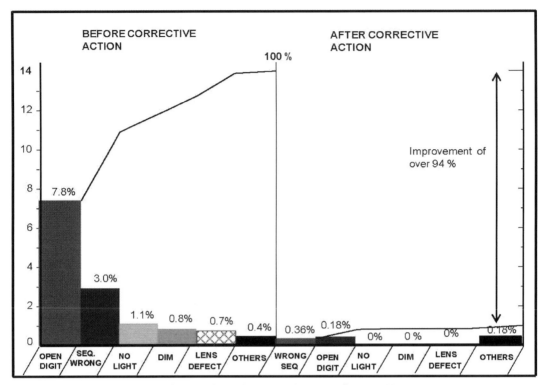

Figure 7-11: Paired Pareto of before and after corrective action

Step 6: Take Appropriate Action

The team implemented the respective corrective actions and fixture and incorporated into the process specification, documentation, and *control plan*. The results were conveyed to all employees and the importance of the new procedures emphasized.

Step 7: Future Plans

The team's future plans are as follows:

1. To continue monitoring of the first pass yield of QDSP-678.
2. To extend the corrective action of this project to all similar assembly type devices in the production line.

Benefits of the PDCA Improvement Cycle and Why It Works

By now some of the benefits and strengths of the PDCA cycle should be clear. Here is a list of the main benefits:

- It is a systematic problem solving process that provides the quickest route to an effective solution.
- It involves a team of people who are close to the problem at hand.
- It ensures an agreed upon schedule for project completion.
- It ensures an agreed upon goal or target, usually set with data.
- It ensures a detailed analysis of the failure modes and current workflow.
- It ensures verification and elimination of the root cause of the problem.
- It requires implementation of controls to monitor and manage the new improved process.
- It requires training in and documentation of the revised standard work.
- It requires documentation of before and after fail data. This will be useful for the next improvement cycle.
- It will ensure no recurrence of the problem, thus ensuring continuous improvement. This is achieved through standardization of the new improved process.
- Managers and supervisors may come and go but if the PDCA process is understood and institutionalized, the result will be: Employees will always be systematic and analytical in their approach to problem solving and continuous improvement.

The last point stated just above is important. We paraphrase it as follows: *People may come and go, but processes must continue.* The PDCA cycle is part of a process – the improvement process. Over the years, we have seen numerous improvement processes, each customized to meet an individual's needs or pet theory. This causes an excessive amount of re-learning and inefficiency. Hence it is important to adopt and standardize to one methodology for all improvements.

Uses of the PDCA Cycle

The PDCA Cycle is shown as a circle or wheel – this implies that we keep going around in a cycle of *continuous of improvement*.

In general, all approaches to problem solving are cyclical attempts to identify the root cause of a problem, eliminate it, review the new status, and repeat the cycle. If the root cause is not eliminated and the problem is not resolved, then we need to go back to the analysis stage and go through the problem solving sequence again. The PDCA cycle or its variants are used in manufacturing, the military, and business in their planning and improvement processes. Here are some examples:

- The 8D process: initiated by the US military and also called the Ford 8D process. This uses the PDCA cycle in 8 detailed steps. This is a standard

format for CARs: Corrective Action Requests from customers or internal quality departments.
- The OODA loop (Observe-Orient-Decide-Act), was developed by military strategist John Boyd. According to this concept, the key to victory is to be able to create situations wherein one can make appropriate decisions more quickly than one's opponent. This is used mainly in military organizations.
- The Six Sigma improvement cycle or "DMAIC project methodology": Define the problem; Measure key aspects of the process; Analyze the data to understand and verify cause & effect relationships; Improve the process; Control the future state.
- Management driven improvements and project management using A3/A4 process. The A3/A4 processes uses the PDCA cycle and is discussed later in this chapter.
- New design process: plan and design the new product; run the first pilot production; check results; improve the design; repeat the cycle until satisfied.
- Hoshin Kanri planning process and reviews: this is discussed in the chapter on planning.
- The TOC (theory of constraints) improvement cycle: Identify the constraint; exploit the constraint; subordinate everything else to the constraint; elevate the constraint, and remove it. The TOC term for root cause is "constraint" or the weakest link. This concept comes from German manufacturing theory of managing takt time, which strives to continuously remove the weakest link of the slowest time in a series of manufacturing steps, in order to remove bottlenecks and speed production. This is an ongoing cycle and as each weak link is removed the operation gets more productive and efficient. The approach of identifying the weakest link sharpens the problem solving process. We will discuss the TOC approach to planning in the chapter on planning.

Seven Quality Control Tools and Other Methodologies

In our discussion on the PDCA cycle, we have mentioned many QC (quality control) tools that could be used in the steps. The seven tools are:
- Data Collection Checksheets and Checklists
- Pareto Diagram
- Cause and Effect or Ishikawa Diagram
- Stratification
- Graphs and Histograms
- Scatter Diagrams

- Control Charts

According to Dr. Ishikawa, about 95 percent of the problems in the work place can be solved using these tools, used within the context of the PDCA cycle. Refer to the appendices for references on the seven tools.

Types of Improvements Projects

Continuous improvement is a never ending process because we strive towards increasing customer satisfaction, reducing waste, lowering costs, and better quality. To quote Dr. Deming: *"We need never ending improvement to establish better economy"*. Improvement projects can be driven by customer requests or complaints. This is the most important reason for doing an improvement project.

Consider the following quote by John Akers – Ex Chairman IBM Company: *"I am sick and tired of visiting plants to hear nothing but great things about quality and cycle time - and then to visit customers who tell me of problems."* An old quote but very relevant today – the need to be close to the customer can never be overemphasized.

Kaizen projects: Kaizen stands for a philosophy of continuous improvement of business and manufacturing processes. However, do note that the kaizen process encourages small and quick improvements in an operation. But to be successful a company needs both kaizen and innovative breakthroughs.

Improvement projects driven by production (kaizen) teams: Often these are a part of the annual company plan or the department's productivity goals. One such project was discussed earlier in this chapter. Projects will include reduction of the seven wastes, for example:

- Overproduction: Moving from batch to one-piece production.
- Unnecessary transportation: Production line improvement from equipment based to product based layouts.
- Unnecessary motion: Reviewing and improving standard work.
- Waiting time for operators/machines: Setup reduction/ SMED projects.
- Excess inventory, kanban, or WIP reduction projects.
- Quality defects: Reducing test fallouts, or inspection defects.

Breakthrough projects can be driven by top-management and cross-functional teams. Often these are a part of the annual company plan and address major, high impact, breakthroughs. This will include Management driven projects shared across an organization or department using the A3/A4 report.

Small day-to-day improvements on the production line or warehouse; these can be quick improvements suggested by operators, technicians, or engineers. This can be a tweak of standard work or production equipment that improves productivity or quality.

126   *Winning with Operational Excellence*

Such projects may not have major productivity improvements but encourage employee engagement and participation; and create a healthy and stable workforce.

The above list gives some options. However, it is crucial that we do not start numerous small unconnected improvement activities throughout the organization, or set a quota for monthly kaizen projects. This can result in islands of productivity within a factory which may create lack of focus or drive wrong priorities. Priority setting and only fixing what is necessary is very important and discussed in more detail in the chapter on getting started. We discuss kaizen teams in detail in the chapter on employee development and participation.

Rate of Improvement and Setting Targets

What is a good rate of improvement? How aggressive should improvement targets be? For many years we have come across several Quality and Kaizen teams that use a rule of thumb of a 50 percent reduction in defects for each project, but we have never been able to trace the source of this rule.

An article, "Setting Quality Goals" by Schneiderman[26], sheds some light on this question. Empirical evidence suggests that most improvements (more specifically reduction in defect levels) can be made at the rate of 50 percent in a very narrow time range of 6 to 9 months. Reductions of defects in an autonomous environment like manufacturing or administration take less time. Reduction of defects in a complex environment such as between a factory and its supplier may take longer. The data shows that the rate of improvement is independent of cumulative volume or the learning curve, but is dependent only on time.

The reason for this phenomenon is simple: Typically a few causes, the first few items on the Pareto diagram of defects, are responsible for a majority of the defects. Hence eliminating the first few items will quickly reduce defects by 50 percent or more. This cycle can be repeated several times. Refer to the PDCA project in this chapter.

As an initial rule of thumb, we would recommend you plan to set a target of 50 percent reduction in defects every 6 months. By defects we mean waste, scrap, inventory levels, time to do something, and any other undesirable item. Once you set a target based on this simple rule you can plan to deviate from it only if data suggests you can do otherwise.

Management Improvements via the A3/A4 Improvement Process

The PDCA cycle provides a powerful format that can be used to manage projects and request for improvements. Many companies use this format to communicate needs for analysis and improvements. As shown in the previous section, the PDCA cycle is a systematic, data driven, and analytical process, which is used to improve or manage products or processes. The A3 improvement process is an important communication and improvement tool at Toyota and other quality driven companies.

Here we will discuss management's use of the PDCA cycle using the A3 problem solving process or report. The term "A3" comes from the paper size used for the format, which is equivalent to B-sized paper. Nowadays, with the increased usage of e-mails and computers, the A4 size (equivalent to letter size) is becoming popular. With the advent of computers, it is now common to embed files and photos as the A4 template is passed around between management and professionals via emails.

There are many variations of this format, for managing projects and improvements. We show here the format used for improvement. Other formats with a different requirement still use the PDCA steps but with room for different data requirements. The important point with the different formats is the systematic approach using the steps of the original PDCA cycle.

Many consultants and companies conduct one to two day seminars or training on the A3 process – basically the bulk of such efforts is a detailed education of the PDCA cycle, enveloped within the A3 format! We have already discussed the PDCA cycle in detail therefore we will discuss the benefits, use, and an example of this methodology.

The A3 report is used as a tool to record a request for improvement, communicate to get relevant inputs from relevant managers and staff, and then to propose the final resolution. Use of such a report by circulation and getting inputs from a team until resolution is reached facilitates:

- Constructive communication and dialogue between all parties
- Knowledge sharing with the team and management
- Teamwork within the organization
- Systematic and data driven improvement
- Good documentation of the entire project for future reference and further improvement in the true spirit of kaizen.

The A3 process when used by managers facilitates the *nemawashi* process, which my friend and mentor, Dr. Noriaki Kano, is very fond of proposing. *Nemawashi* is Japanese for an informal communication process between the management team, which helps to set the foundation for changes or for introducing new projects. The procedure in a Western company would be to make the proposal at a formal meeting, but the nemawashi process is indirect and the proposal is first shared so that inputs and changes are collected until a resolution is reached. Only then is the formal proposal released and it typically gets quick approval. The A3 process can be used to follow a similar process, via emails, to ensure a proposal gets full consensus before implementation.

The A3/A4 format is shown in Fig. 7-12. Note that it follows all the steps in the PDCA cycle. The A3 report must:

- Be concise and to the point. The steps of the PDCA cycle are followed, however, there is room for the main points and a good summary is essential for the report to communicate the issues at hand.
- Be visual: Wherever possible add diagrams or photos to convey your information.
- Effectively communicate the problem at hand.
- Be shared and distributed to the right parties and allow them to review and provide solutions and suggestions for improvement.
- Be completed with the resolution in a reasonable timeframe.

Managers should be encouraged to use this A3/A4 process whenever they need to plan for changes and improvements. The use of the PDCA cycle will help them use facts, data, and ensure a good analysis to drive towards a solution. Furthermore this process will help employees, teams, and managers to engage in teamwork, and analytical problem-solving.

A Case Study using the A3/A4 Improvement Process

We share a case study from manufacturing using the A4 format. The genesis of this project was a concern by the quality manager regarding congestion, quality issues, and defects on the production floor. Therefore an A4 report was prepared and sent to several managers to get inputs. The report was also routed to engineering to get a detailed proposal. We show the completed report in Figs. 7-13 and 7-14.

Fig. 7-13 (the top-half of the A4) shows the analysis, proposal, and improvement targets. After much deliberation, meetings, e-mails, and other communications the solutions were proposed.

The final completed report is shown in Fig. 7-14 (the bottom half of the A4). The completed report shows the counter-measures taken and the results achieved. The results include cycle time and defect reduction. Photos are embedded in the report to show some of the improvements.

Quality: Improvement and Control

A3/A4 IMPROVEMENT PROCESS

0-BACKGROUND -EXPLAIN ISSUE/PROBLEM NEEDING IMPROVEMENT	Project Originator/Team	
Give a short background of the issue or problem - what difficulty is it causing?	Date:	
	Requested Finish Date:	
	Actual Close Date:	

1. PROJECT THEME/OBJECTIVE

State your objective - What do you want to achieve?

2. GRASP & EXPLAIN THE CURRRENT STATUS - WHAT ARE THE ISSUES - USE DIAGRAMS, PICTORIALS, FLOW CHARTS

3. IDENTFY PROBLEMS, ISSUES, ROOT CAUSE, VERIFY IF REQUIRED: Use 5 Whys, Cause & Effect Diagram, and illustrate your problem

4. STATE COUNTERMEASURES REQUIRED & DETAILED TARGET TO BE IMPLEMENTED: Use diagrams, pictorials to support your request

State the Specific Countermeasures that can address the root cause.:

What are the changes to be made that can move the organization to a better state?

Detailed Targets

What will the the new state look like after the changes have been made? What must be achieved?

5. DESCRIBE IMPLEMENTATION PLAN; Identify Potential Barriers and resolution: GET COST & PROJECT APPROVAL			TARGET DATE & COST	
Describe the implementation steps in detail	ROADBLOCK	RESOLUTION	Target date	
			Estimated Cost	
			Cost Approval	
			Project Approval	

6. CHECK THE RESULTS OF THE IMPLEMENTATION PLAN

Check the resuts of the implementation plan. Has the implementation plan been rolled out as planned and results achieved? If not, why not?: You may need to go back to step 3 & 4, and resolve.

7. STANDARDIZE THE NEW PROCESS - TRAIN, DOCUMENT, & COMMUNICATE	Responsible Person:	

Figure 7-12: A3/A4 Improvement Process Report

A3/A4 IMPROVEMENT PROCESS				
0-BACKGROUND -EXPLAIN ISSUE/PROBLEM NEEDING IMPROVEMENT			Project Originator/Team	Soin Singh
Currently there is too much physical movement of Customer products (pcb assemblies and finished goods); label, packaging, and FGI stations are congested together.; furthermore storage of WIP awaiting final configuration is along the aisle leading into production line.			Date:	16-Mar-11
^			Requested Finish Date:	10 May 2011
^			Actual Close Date:	
1. PROJECT THEME/OBJECTIVE				
Reduce physical handling & transport of Customer products, using lean manufacturing principles, with goal of reducing physical damage and label/print defects by 70%.				
2. GRASP & EXPLAIN THE CURRRENT STATUS - WHAT ARE THE ISSUES - USE DIAGRAMS, PICTORIALS, FLOW CHARTS				
1. Current line is short and lacks space for additional IPQC and QA Out of Box Audit (OBA) 2. Sharing of label equipment create cross-traffic 3. Therwe is excessive inventory of packaging material, boxes, pallets -- refer to attached layout and photo. 4. Packaging process creates too much debris in production area 4. Labeling/packaging errors are high, rejects at customer QA are high				

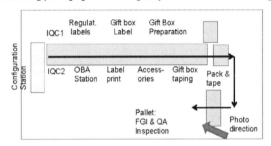 

3. IDENTFY PROBLEMS, ISSUES, ROOT CAUSE, VERIFY IF REQUIRED: Use 5 Whys, Cause & Effect Diagram, and illustrate your problem
Congestion is due to: Line converted from low volume prototype line to high volume in same space/footprint. Issues: a) **Equipment shortage:** Too much operator cross-movement & travel in work area due to one of each equipment (print & dispense). b) **Congestion:** on line at IPQC and OBA station, hinders ability to do effective inspection c) **Lack of Kanban process:** Restricted space/process for gift-box & carton preparation, causing congestion and excess inventory. d) **Scanner errors:** due to poor label printing (reject at final QA scan before shipment)

4. STATE COUNTERMEASURES REQUIRED & DETAILED TARGET TO BE IMPLEMENTED - Use diagrams, pictorials to support your request	
State the Specific Countermeasures that can address the root cause.:	
Equipment Shortage	Acquire duplicate equipment, implement parallel processes, plus add extra equipment/workstation
Congestion	Convert straight line to cellular production line. Provide space for 10-15 workstations, based on product mix
Kanban Process	Install 2 workstations for gift-box and carton production. Volume to be controlled by signal Kanban process.
Scanner Errors	Purchase better scanners plus review and improvement print quality, also review scan software.
Detailed Targets	
a) Reduce Kanban box inventory to max 30 min quantity b) No visable congestion c) Operator production process to meet ergonomic standards d) Reduce defect rate by 80%	

Figure 7-13: A3/A4 Improvement Process With A Proposal

5. DESCRIBE IMPLEMENTATION PLAN; ID Potential Barriers and resolution. —GET COST & PROJECT APPROVAL			TARGET DATE & COST	
Describe the implementation in detail	Roadblock	Resolution	Target Date	
Proposed layout is cellular and shown here: 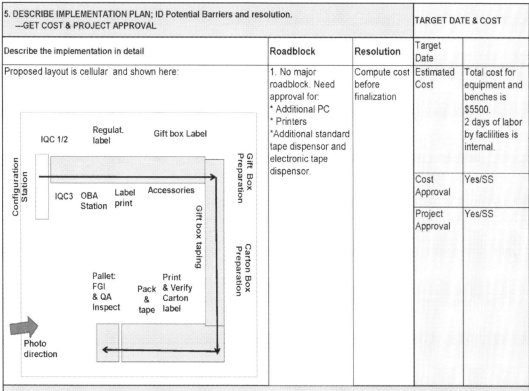	1. No major roadblock. Need approval for: * Additional PC * Printers *Additional standard tape dispenser and electronic tape dispenser.	Compute cost before finalization	Estimated Cost	Total cost for equipment and benches is $5500. 2 days of labor by faclilities is internal.
			Cost Approval	Yes/SS
			Project Approval	Yes/SS

6. CHECK THE RESULTS OF THE IMPLEMENTATION PLAN

Final implementation has resulted in smooth continuous flow operation in a U-Shaped Cell.
There is no congestion of material and opertors.

Performance measures show improvement as follows:
*Cycle time has improved by 3 secs
*Label and pacakaging rejects have decreased by 60%.
*Refer to embedded peformance measures chart and photos.

Responsible Person: Azziz

Due to new introduction of new processes, the majority of the work standards have been revised. Opertors have been trained in the new standards. QA will audit for compliance.

Figure 7-14: Completed Project Report Using A3 Process

Problem-Solving Hierarchy

Problems and errors will occur. Our goal then is to minimize problem and error occurrence in the organization. We will discuss further on how to prevent human errors later in this chapter. Meantime, as we eliminate problems and decrease errors, we will go through the following phases or hierarchy:

1. *A Fire Fighting Mode.* Here problems occur helter-skelter throughout the organization and we fix them on an ad hoc, non-systematic, basis. Everything is disorganized: from defects in production, shortage of raw materials, to late deliveries. Often, the bearer of bad news is punished. The person who fixes the problem and douses the fire is rewarded and considered a hero; but no attempt is made to understand why the fire occurred, or why there are so many fires.

2. *Systematic Problem Solving Mode.* Here we are in control, fixing problems systematically, eliminating root cause, and operating in an environment of continuous improvement. The PDCA cycle is used constantly and consistently; systems are in place to detect problems and resolve them quickly. As a result defects are low and most performance indicators are showing continuous improvement.

3. *Prevention of Problems by Prediction Mode.* Here we go beyond systematic problem solving; instead we *predict and prevent* potential problems from occurring. In this stage we design and use good processes to reduce waste and defects: An organized operation using the 5S system, use of meticulous and effective work standards, managing cycle time on the factory floor, minimizing WIP, and speeding throughput in manufacturing via continuous flow.

4. We are able to understand customer and quality needs before we design a product or service; hence products and services will have "attractive" quality features. Most potential design and process problems can be detected and prevented with the use of DFM (design for manufacturing) and Process-FMEA techniques; this will ensure customer acceptance, less design changes, and less defects and fallouts in manufacturing.

5. In this stage we move into the TQC stage – Total Quality Creation; we plan for and ensure total customer satisfaction. This third stage is crucial if we wish to move towards operational excellence and World-Class manufacturing. Clearly, this is the phase we want to be in – always.

Relationship between Improvement and Control

The PDCA Cycle is shown as a circle or wheel – this implies that we keep going around in a process of *continuous of improvement*. However, there are times when we wish to defer continuous improvement and divert resources to other areas and challenges. At that point, we need to ensure that we leave the current improved process in a state of good, tight, control, with no recurrence of issues. This requires completion of the Act step of the PDCA cycle.

This concept of selecting between further improvement and control is shown in Fig. 7-15. The choice is influenced by the status of the current project and other priorities. The purpose of putting it under control is to maintain the improvements that have been made, in order to ensure we do not slip into old habits and lose the gains. At the end of each project, we must:

- Update standard work procedures plus re-train production staff.
- Ensure good documentation of completed projects: This provides a reference for the next improvement cycle.
- Provide recognition and reward to kaizen teams.
- Track and manage the improved process with yields and control charts.
- Update the previous Process Management Plan (PMP) to reflect all changes and improvements.
- Conduct an audit to monitor process stability of the improved process.

In the next few sections, we discuss some of these methodologies.

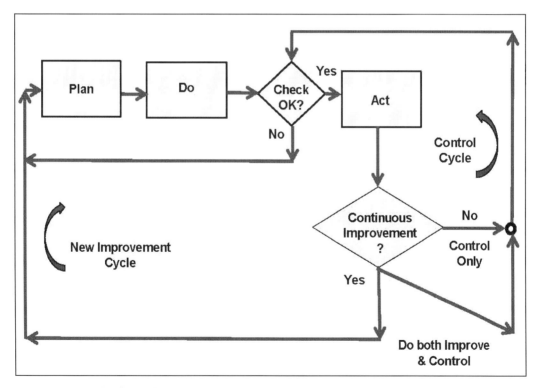

Figure 7-15: The relationship between control and improvement

Quality Control and Management

The Need for Process and Quality Control

We have just discussed the need for control after an improvement project is completed. In fact all processes within an operation need to be in a state of control. Here is an anecdote that illustrates the need for good process and quality control[27]:

My friend and mentor Dr. Noriaki Kano, a member of the Deming prize committee in Japan, was once reviewing a manufacturing operation. He had been impressed with everything the managers had shown him. As we passed a production operator working on a bonding machine, he stopped to talk to her.

He asked her if she knew the reject rate of her bonding machine. "Oh yes, she replied, it is 0.15 percent." What did she do if for any reason the rejects exceeded that number? She replied that she had been told to call the supervisor whenever rejects exceeded 0.20%. Dr. Kano decided to check the records to verify her statement. He flipped through her records and found several days when the lot reject percentage was 0.25%; so he asked her "What did you do?" She replied that 0.25 was so close to 0.20, that she did nothing. "Fair enough" he said and looked at the records again and found another lot about one week before with a reject rate of 0.30%. "What did you do then?" he asked. She replied that she had indeed informed her supervisor.

The supervisor happened to be standing by and informed us that she had called the engineer. Dr. Kano remarked, "It's been a week, what has the engineer done?" The engineer was nowhere to be seen. Dr. Kano suggested we look for the engineer and sent someone to find him. Meantime, the production manager was getting a little nervous and suggested we move on. Dr. Kano stood his ground and asked to meet the engineer. We finally found the engineer. He had been informed but what had he done? He had done nothing, no documentation or analysis was available; he just never got around to doing anything, but he planned to do something – when he had time.

The learning points of this episode are:

1. Engineering had done a good job of characterizing the process per theory.
2. When the process went out of control, the operation did not have the discipline to analyze and take corrective action.
3. Eventually the process could have given more rejects, gone out of control, but nobody would have noticed – until it was too late.

After this experience, Dr. Kano commented, *"The workers are awake, the engineers are just waking up, but management is asleep."* His point was that the workers were doing their best and alerting management to potential problems; the engineers were busy and stretched; but management was oblivious of the real problems. In this case, there was a possibility that problems would multiply and the process run out of control. The outcome could have been product problems and customer dissatisfaction.

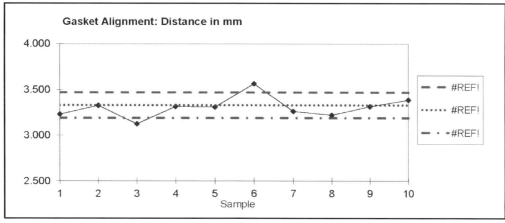

Figure 7-16: SQC Chart for Gasket Alignment

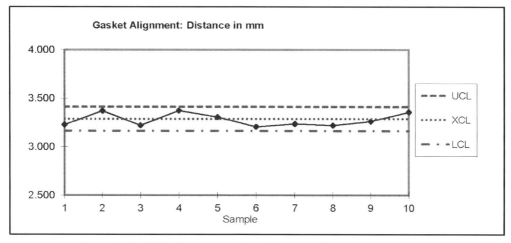

Figure 7-17: SQC Chart for gasket alignment after process improvement

Statistical Process Control

Statistical process control (SPC) is a vital tool to help control and monitor quality of processes. SPC, also termed statistical quality control (SQC) is a well established tool and training and many texts are readily available. Therefore we will give a short write-up of the benefits of SPC here and expect the reader to go to the appendices for more information.

The basic requirement in a manufacturing process is to establish a state of control and sustain this state through time. A control chart enables us to observe if a manufacturing process is working correctly and "in control".

A control chart has a center line corresponding to the average quality at which the process should perform. If statistical control of the process exists, there will be two control limits: Upper Control Limit (UCL) and a Lower Control Limit (LCL). The average (XCL) and the two control limits are computed from the data. The process performance is then charted. An out of control situation, trends, and other unnatural patterns can be easily detected from the chart. When abnormalities are detected the production team will take corrective action to bring the process back into control.

We demonstrate the essence of control charts with the following example: In Fig. 7-16 we show the SPC control chart at a workstation. An operator at a workstation places a rubber gasket on to a hand-held computer touch-screen; after the gasket is placed onto the touch-screen, the screen is assembled into the computer cover. It's important that the assembly process conforms to specifications and that the gasket is placed at the correct position – if it's too close to the active boundary of the touch-screen it will malfunction, but if it is far away and too close to the edge there is a visible gap. The gasket alignment, or distance from the active boundary of the touch-screen at a critical point, is measured and the data plotted and monitored on a control chart. The data is collected hourly on a sample basis; the control chart is plotted and monitored to ensure the process is in control.

Fig. 7-16 shows the average alignment – or distance – from the active boundary of the touch-screen after assembly. The average distance is not in a state of control, as is evident from the alignment exceeding the two control limits on several occasions. On investigation, the engineer identified several causes:

- Operator skills in using the gasket assembly-tool.
- The accuracy and effectiveness of the gasket assembly tool.

It was discovered that operator training and skills had an impact. Several different operators had results ranging from good to unacceptable. Hence the first step was to re-train the operators and certify only those who could do the work consistently well. Next, it was decided to improve the assembly tool so that it was less dependent on operator skills.

After the assembly-tool was improved via a redesign, the out of control situation was eliminated. Furthermore, more consistency was shown since the control limits were narrower and closer together. The improved assembly-tool improved the assembly process and eliminated rejects. Refer to Fig. 7-17.

The improvement came after both management and engineer intervened to analyze and improve the process.

Types of Control Charts and Rules

There are several different types of control charts; each is used for measuring different attributes. For example x, R charts are used for measurements (weight, volume, etc.) and p charts are used for number of defects. There are many rules to help the engineer to detect an out of control situation or a potentially deteriorating situation. A manufacturing operation must use control charts to monitor critical dimensions of fabricated parts or track critical process points. What is essential, however, is not the number of control charts in an operation but how they are used to track critical processes. More important, there should be continuous improvement as evident from the charts over a period of time. Refer to the appendices for references.

Monitoring and Controlling Quality on the Line via Audits

Once a process is running smoothly it is necessary to monitor and ensure it is in control and meeting established guidelines. But, why would a process go out of control? There are many reasons:

- Improper standard work, causing errors. Work and training cannot be effective if standard work is difficult to put into practice.
- Errors due to inadequate training of operators.
- Operator errors due to operator lapses or operator not able to keep up with station cycle time, and hence skipping some work steps.
- An increase in production schedule and forcing a faster cycle time in work stations, resulting in operator inability to keep pace. This happens.
- Defective material causing assembly issues.
- Defective or poorly maintained equipment causing assembly errors.
- Poor work flow causing inventory congestion and production errors.
- In fact, a study at Toyota by Spear and Bowen[28] determined that the following failures caused most of their current problems: Improper standard work; weak customer-supplier communication and lack of good workflow for every product and service.

Such issues will occur and recur; consequently it is necessary to conduct process audits to look for them. The objective is not to find fault but rather to surface problems and work with the production and engineering teams on continuous improvement. The entire operation team needs to work together to benefit from such audits.

Therefore, it is important to conduct In Process Quality Audits (IPQA) routinely as a preventive measure and to keep production staff on their toes. The IPQA system formalizes, prioritizes, and schedules the routine checks that must be made in any operation. It is a simple and flexible methodology to ensure that processes are running smoothly in an operation. These audits provide a cost-effective insurance against complacency. Items to check include:

- **Standard work Vs actual operator work**: Here we check for deviations, problem with standard work instructions, or work requiring improvement. Ideally, an effective operator will highlight difficult standard work, but under work pressure and a fast-paced environment this may not happen. This check is done through observations at each work station in production and also reviewing inputs from operators. This procedure is similar to the *genchi genbutsu* and Ohno circle approaches that we discussed earlier. The IPQA audit methodology formalizes these approaches.

- **Practice of 5S, clutter, visual problems, and environmental issues**: Here we check for issues that make it difficult for operators to perform their jobs. This provides an opportunity to review the effectiveness of the 5S system.
- **Process controls charts and machine settings** to ensure machines are running within specified limits. Here we check to ensure machines are running correctly, procedures are followed, settings are correct, and the machine is delivering output to expectations. Typically we check the settings, programs, and other machine requirements – some of which may be monitored on SPC charts. Poor process controls or degrading machine performance may be indicative of ineffective TPM.
- **TPM issues, potential machine problems, leaks, or malfunctions** that have not been addressed: Every operation should have a well established TPM effort; nevertheless, the IPQA provides a check to ensure TPM is working well. Here we review the daily, weekly, and monthly TPM checklist to see if it has been completed.

The IPQA team, engineering, and production should meet and prepare the IPQA procedure flow and checklist. They must define major and minor non-conformance defects. Typically, a major non-conformance is one where the customer will be impacted or the product's reliability is at risk. A minor non-conformance is one where the standard has been violated but causes no impact to the product. In addition, they need to agree on the criteria to stop the production line. An IPQA process flowchart is shown in Fig. 7-18.

Fig. 7-18 shows a recommended IPQA flow and escalation path: If a major defect is found by IPQA inspectors or technicians, the production line must be stopped and the issue resolved within 30 minutes. Alternatively the production engineer can establish countermeasures until the root cause is eliminated – at that point the production line can resume production.

Fig. 7-19 shows an extract of an IPQA report. This audit was done on a production line during the night shift. The issues discovered during this audit are typical. Often, the original standard work prepared at manufacturing release still needs to be improved and fine-tuned – this and other problems can be discovered during the audit. In this instance, because more than one minor non-conformance was discovered (per the guideline in Fig. 7-18) the line was stopped until the fix was initiated.

The IPQA team must be trained to do their work effectively, and they must understand what to look for and when to escalate to stop the production line.

Quality: Improvement and Control   139

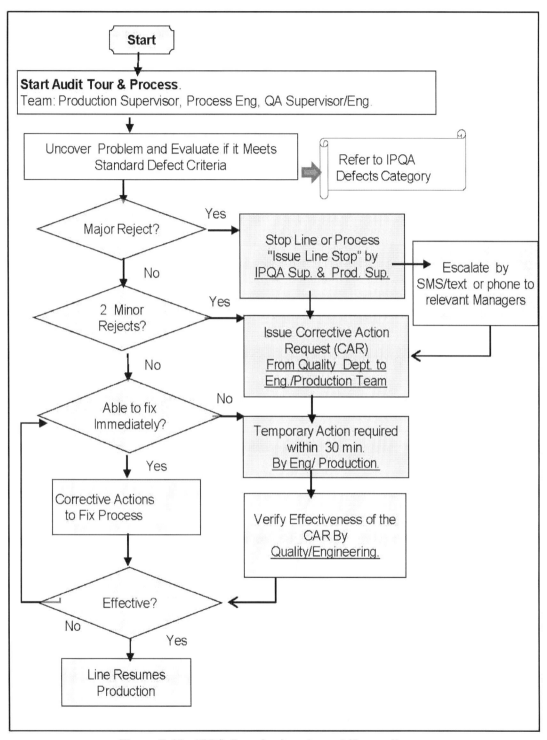

Figure 7-18: IPQA flowchart and escalation path

IPQA Report			Lead Inspector: Frank			Production Supervisor: John		
Production Line: Motor			Shift: Night		Date:			
LINE	Audit Time	Station	Non-Conformance: Major or Minor	Type	Check By	Owner		Status
Motor Main Assy.	9.01pm	AA-02	**Operator Skipped Process:** Operator did not slide or move the width adjuster assembly to right and left to check for smooth non-frictional movement and no any squeaking noise sound. Also did not move length adjuster upward and backward to ensure smooth movement. Operator Name Genn. Badge No. 456	Minor	John	Prod. Sup.		Closed
	9.15pm	AA-05	**Material Issue:** Operator should use cable tie on wire harness which is pre-measured and marked at 50mm interval. But operator estimated cable tie position, because wire harness came with no marking. Supervisor had given her approval to do this due to schedule requirements. Operator Name: Jen. Badge No 0674	Minor	John	Purch. Engr	Open	Short term fix
Motor: Sub-Assy	9.35pm	SA-06	**Operator Skipped Process:** Operator did not follow Work Sequence: She applied the oil applicator jig towards the left only but she is required to move the oil applicator jig towards the left and right end of shaft. Operator Name: Bimi, Badge No 237	Minor	Harry	Prod Sup.		Closed
	9.40pm	SA-07	**Kanban Bin Mixed Up:** We found parts mixed together inside one bin (Spoke - 3mm pick and Hub-3mm Tire). This bin had no part number. Operator Name: Ana, Badge No 129	Minor	Harry	AK -Delivery	Open	Short term fix

Figure 7-19: Extract of an IPQA report

Aiming for Zero Defects

Aiming for zero defects is an approach that many Japanese companies take seriously. Many Western companies, however, prefer to use the Six Sigma approach because the concept of "zero defects", or total perfection, is considered impossible; what's more, the Six Sigma goal is crisp, clear, and easy to comprehend.

The Six-Sigma approach is to gradually improve quality until a defect rate of Six-Sigma or 3.4 defects per million "opportunities" is achieved, with cost savings. The Six-Sigma approach also uses the PDCA cycle to structure the improvement process. We prefer the original PDCA cycle plus the seven tools for the majority of the improvement projects; but complex projects will require statistical techniques similar to Six-Sigma.

Our approach, however, is *aiming for zero defects,* because total perfection is difficult to reach. Zero Defects is theoretically possible but practically expensive, because the preventative cost would be too high.

To aim for zero defects, we need to agree on an overall strategy: Good design is the precursor to high quality products with low defects. Most of this effort will come from the R&D and engineering team. However, once the product is in manufacturing, it behooves us to ensure that we have impeccable processes to ensure the highest quality with the lowest errors and defects. From the beginning of this text we have discussed activities that will give us this goal; from 5S to standard work to machine maintenance and to various strategies to reduce waste. Here we look at four specific activities that can help to reduce errors and move us towards zero defects. These are:

1. Source or Incoming Inspection of parts.
2. Preventing Human errors through Poka Yoke.
3. The next process is my customer and successive inspections.
4. Jidoka and Andon Systems

Source or Incoming Quality Inspection

Prior to product manufacture and assembly, we need to ensure we receive first-rate purchased parts from our suppliers. These parts must meet our specifications in form, fit, and function. Hence we need a well documented inspection procedure, and all parts need to be inspected to ensure that they meet our standards. Furthermore, defects occurring early in the stage cost more to fix later on. According to the 10X rule, they cost 10 times more to fix later in the process.

The 10X Rule

The 10X rule[29] states: *A defect created early in the production stage but detected later down the production chain incurs a factor of 10 in the cost of repair.* This applies to both hardware and software defects.

Purchased parts or production assemblies are often scrapped, while finished products are reworked. Obviously, it is cheaper to catch a defect early in the manufacturing stage than later in finished goods. However, the cost can be higher than 10X especially if the problem is found by the customer – often field repair cost can be 200% of the product manufacturing cost (as opposed to the defective part's cost) as the product is exchanged and not repaired. Therefore it is imperative we catch and eliminate defective material early in the process.

Methodologies for Early Detection of Defects

Incoming inspection, using a sampling plan is one option. We recommend that we do this for *all* purchased parts. Once we gain confidence we can look at other options such as reduced sampling plans, and supplier inspection reports. Better still, however, is source inspection by our inspectors, this will allow us to move parts from dock to stock.

Next, we discuss and show how this can be accomplished. As always, we need to have a goal of having low defect purchased parts. Some suppliers have the ability to do this, others may need help. For those that need help, we need to work with them on detection, improvement, and preventive processes.

1. **Detection methodology:**
    a. Initiate source or incoming inspection: It's best to have 100% inspection of incoming parts, but this is expensive and tedious, hence we prefer a sampling inspection plan. For electronic and electrical parts, check packaging, specifications, physical condition, and quantity. High volume electrical and electronic parts are often stable after good yields have been achieved by the manufacturer. For mechanical parts: Sample inspect all cosmetic and critical dimensions, also compute Cpk of critical dimensions. This is done by calculating a running Cpk, computed from samples drawn from the most recent inspections.
    b. Set up a quality index to review, monitor, and rank each supplier.
    c. Conduct regular process audits at supplier for critical or problem parts. This supplements incoming inspection but the advantage is that we now move further upstream to detect and prevent defects at the source.
    d. Request supplier to provide production yield reports Vs targets.
2. **Continuous Improvement methodology:**
    a. Review supplier's production yields and improvement targets.
    b. Work with supplier to improve their yields and process capability.
3. **Preventive methodology:**
    a. Monitor the supplier's yields and processes via SPC Methodology.
    b. Review the supplier's machine Cpk for all critical equipment.

The above methodologies can drive suppliers to produce better quality parts or sub-assemblies. Let us look at a case study of an Asian Manufacturer of a copier type product. The company had a new product with quality problems and aimed for low to zero defects prior to shipping its products. To move towards this goal, it started by working with suppliers to improve all purchased parts.

Purchased part quality improvement goals were set based on design or required specifications. Initially the focus was on complex parts or high risk suppliers. These were primarily suppliers of plastic and metal parts; they were given aggressive goals for high risk and critical parts. The manufacturer worked with the suppliers per the 3-step recommendations given above.

Fig. 7-20A shows the results of work done with the first supplier Myoshi, a supplier of plastic case parts. The initial effort gave quick results as the company and supplier identified the major defects and fixed them; these were primarily cosmetic rejects due to handling and molding problems. After that, it was fine tuning of the process until the goal of 100 DPM was achieved. With this background and knowledge, the team worked with other supplies. Fig. 7-20B shows the monthly supplier performance for all purchased critical parts (plastic and sheet-metal). All suppliers are within the target range of 100-500 DPM (defects per million), except for Senko. The next step is to work with Senko.

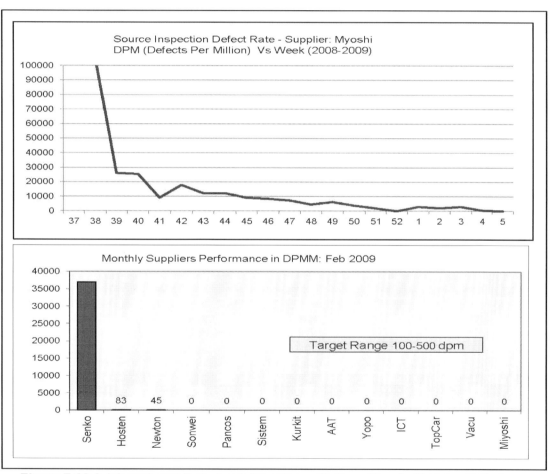

Figure 7-20 A (above) and B (below): Working with Suppliers to Reduce Defects

Preventing Human Errors through Poka Yoke

Before we plan to reduce human errors, which can cause defects, we need to understand the sources of errors. Why are humans so frail that they cannot do it right? What causes errors and defects? *"The root cause is incomplete knowledge or imperfect work"*, says Hitoshi Kume[30]. This causes problems everywhere, in every company, every organization, and with individuals. Errors and defects will always occur. It is our job to minimize their occurrence. These can be reduced with good product design, well designed processes, and superior standard work. Let's look at how we can reduce imperfect work or human errors on the production line.

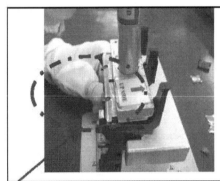

Previous Process:
Operator uses hand pressure & judgment to assemble a flexible cable; too much pressure can cause dimples and damage the cable.

Poka Yoke solution:
A Hand operated press is used; the press movement is limited by design to prevent over-pressure or damage.

Figure 7-21A: Machine or Equipment Limit Poka Yoke

Picture shows an SMT machine which loads components on pc boards. In cases of miss-loads or dropped components the machine shuts down, turns on an Andon light, and a LCD display will display and identify the error.

Figure 7-21B: Machine Shutdown Poka Yoke

Poka Yoke

Human errors can be avoided with the introduction of *poka yoke*. Shigeo Shingo[31] of Toyota Company coined the term *poka yoke* which means *mistake-proofing* or *fail-safe*. The purpose is to eliminate defects caused by human error.

Implementing poka-yoke is an effective way to reduce human error and gets towards the goal of less or zero defects. A good example of poka yoke which we experience every day is a 3-pin power cord: One end goes into an appliance and the other end into the powered wall outlet. Because of the way the plugs are designed, a user cannot make a mistake and plug it in wrongly. Furthermore the plugs and sockets for 110 volt operation are different from those for 230 volt operation; this again prevents mistakes from happening. On the other hand the various cables connected at the rear of a desktop PC Computer can be confusing!

The poka yoke process can be implemented within a very wide range of options. These range from bullet-proof (machine forced) quality to simple check lists. There are four broad categories of poka yoke:

1. **Machine or Equipment Limited**: The equipment or machine ensures, or forces, the exact correct work to be done. See the example in Fig. 7-21A. Here we have a tool or machine with set limits to prevent errors. Or it can be plugs or contacts that fit one way only. In this mode, production never stops for defects

2. **Machine shutdown:** In this case the machine shuts down if an error is detected. An example is an SMT machine that loads components on PC boards: If it detects that a component is not loaded properly, it stops operation. Fig. 7-21B, shows an SMT machine that has shutdown because of a component miss-load. A warning light plus computer generated readout displays the details of the error. Typically, such machines come with the special shutdown feature. This method is similar to the jidoka system or smart machines – discussed later.

3. **Warning or Alert type:** A warning is given if an error is made, and the operator has to redo the sequence of actions. If the error persists, the operator may not be following instructions or the procedure may be incorrect; in that case she needs to call the supervisor or engineer to resolve. See Fig. 7-21C.

4. **Performance Checklist**: Here we use a checklist to ensure the operator goes through a list of actions. See example in Fig. 7-21D. Some daily examples of a checklist: A shopping checklist or the checklist that provides a pilot with a sequence of actions to do before take-off or landing.

146   *Winning with Operational Excellence*

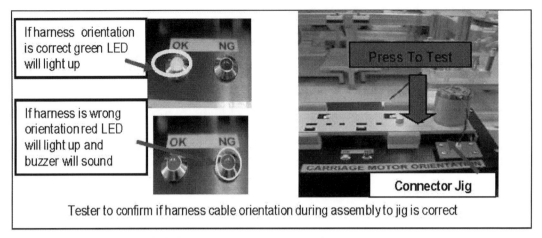

Figure 7-21C: Warning or Alert Poka Yoke

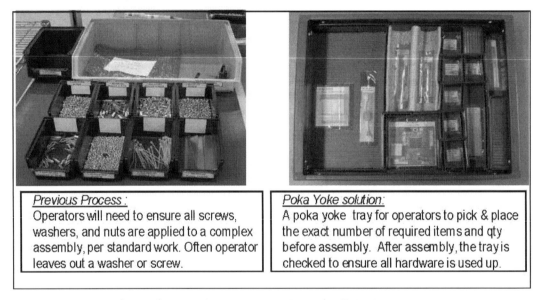

Figure 7-21D: Performance Checklist Poka Yoke

This list of the various types of poka yoke is in the order of most effective (machine limited) to least effective (performance checklist). The first, machine limited, leaves out human judgment and is solely dependent on the machine design. The second, machine shutdown causes the machine to shutdown and provides failure information to the operator when an error occurs.

The third, warning or alert type, reminds the operator that a failure has occurred and further action is required by the operator. The fourth, performance checklist, is the weakest and if the list of actions or instructions is not followed, failure can occur. Warning and performance checklists are the weakest poka yoke methods as they are operator dependent.

The poka yokes that are implemented should be simple, easy to use, and inexpensive. The examples quoted here fit these requirements. Well-designed *poka yoke* systems are akin to doing 100% inspection without the tedium of doing operator assisted manual inspection. Hence *poka yoke* is desirable, necessary, and should be implemented as much as possible in the manufacturing environment.

The Next Process Is My Customer

One of the important tenets of Total Quality Management is *"the next process is my customer"*. This means that I must make a product that is good for the next person, who is my customer. This concept ties in with one of the rules of the kanban system which we discussed earlier: *All products manufactured and sent to the next process must be 100% defect free.* This requires employees to ensure they only pass good quality reports, products, assemblies, and service to the person within the company – that next person is the customer and must be treated as such. Similarly, operators who receive any material from the previous process must inspect them.

Successive Inspection System

Production operators should be inspecting assemblies they receive from the previous process: This is called a *successive inspection system*. This is especially crucial in operations after a very critical step is implemented on the assembly line. This procedure should be used as often as possible – this will facilitate the move towards zero defects. An example of this is shown in Fig. 7-22, which is an Inspection, or Process and Quality Alert, notice to the relevant operator. In the figure, the operator has to check that a motor harness, from the previous process, is tightly plugged in. This procedure forces 100% inspection for critical areas, that are difficult to control but which must be defect free, so that problems do not occur later in the operation or during product use at the customer site.

Jidoka and Andon Systems

The principle of *Jidoka* describes machine capability that has intelligent design to stop a manufacturing process whenever abnormalities occur. This was popularized by Shigeo Shingo, of Toyota, who coined the term *autonomation* to describe this process in English. There are numerous stages in manufacturing ranging from purely manual to fully automated work. Ideally, machines must be fully automated to detect defects and correct their own operating problems but this is not cost-effective. However, most of the benefits of full automation can be gained by autonomation or Jidoka. To be effective, the Jidoka system must be ruthlessly enforced. One example of this is shown in Fig. 7-21B.

148   *Winning with Operational Excellence*

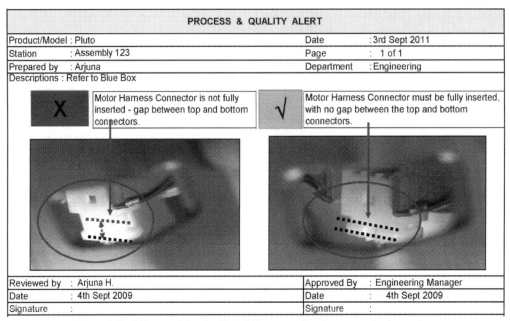

Figure 7-22: A Process & Quality Alert Notice, requesting for successive inspection

An effective Jidoka system takes over the supervisory function of running machines. This means that if an abnormal situation occurs, the machine stops and the operator will stop production. This relieves the operator of the need to continuously judge or measure whether the operation of the machine is normal or defective. The Jidoka system can also be manually operated by the production operator if the operator detects a defect – in this case the operator turns on the Jidoka and stops the line and a supervisor or technician comes in to help resolve the issue. Refer to Fig. 7-21B for such a system. The operator's efforts are now focused on producing good parts. Hence Jidoka prevents the production of defective products and directs attention to resolving the problem at hand. Specifically the jidoka system:

1. Detects the abnormality – automatically or via the operator.
2. Stops the production line or machine: The line supervisor or his/her designate will work with the machine operator to resolve the issue.
3. The operator fixes or corrects the immediate condition, by implementing a countermeasure.
4. After a countermeasure is taken production restarts and a more permanent solution has to be found – unless the engineer considers the initial countermeasure as the permanent solution.

Ideally, of course, Jidoka system follows the PDCA cycle and requires that we investigate the root cause of the abnormality and implementation of a long term preventive measure. However, this may or not be possible, hence if initially a countermeasure is taken, then we have some time to investigate and ensure resolution. This allows the production line or machine to restart and manufacture, while the preventive action is considered. If no quick countermeasure is possible, then the supervisor has to call in more expertise to resolve the issue.

To complement the Jidoka system, after the defect is discovered, an Andon (or light) display is triggered. The Jidoka and Andon system can be of several types:

- A failure detected by the jidoka system will trigger a flashing red light plus an alarm to alert the operator or technician that a defect has been discovered. Note: *andon* is Japanese for lantern or light.
- A display panel that indicates which among several machines has defects. The system typically indicates where the alert was generated, and may also provide a description of the trouble.
- Many modern machines have a Jidoka system that provides a LCD display, with graphics or text, to indicate the nature of the issue and defect. This is common with SMT and Printed Circuit Board manufacturing equipment, shown in Fig. 21B.
- The Jidoka system may include a means to stop production so the issue can be corrected.

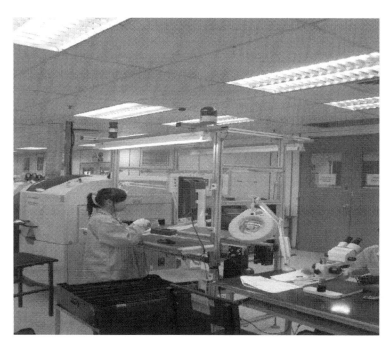

Figure 7-23: Andon system with warning light and alarm.
Here the operator is warned of a failure at an AOI (automatic optical inspection) station. When she sees more than 3 similar defects she triggers the Andon alarm.

The alert can be also activated manually by a worker flipping a switch, after the machine displays a defect. This type of Jidoka and Andon combination is implemented where a machine or production line does not have an automated stop-production process. Refer to Fig. 7-23 for such a process. Here the operator is warned of a failure at an AOI (automatic optical inspection) station. When she sees more than 3 similar defects within a given timeframe, she triggers the Andon alarm, which has a warning light and audible alarm. A technician needs to come to the station, switch off the alarm and attend to the defect: In this case, check the equipment generating the defect, and come up with a short-term and long-term fix.

Setting Quality Goals

What is a good guideline for setting quality goals for processes? For a start we need internal and external customer oriented goals. Customer oriented goals will measure those items that the customer will be experiencing. However, all the customer goals, such as on-time delivery, reliability, and quality, will be driven or influenced by internal processes. These customer goals will depend on processes; hence the goals set internally have to ensure that these external goals can be achieved. Therefore we need a guideline for internal goals. Our guideline is influenced by our philosophy: *A good process will give good results, whereas a poor process will yield poor results.*

Therefore:

1. Identify the critical processes in the operation.
2. Set performance metrics for these critical processes: We can either plan for improvement or control of the process.
3. Track routine performance of these processes.
4. If the performance metric is below target, intervene immediately, and get the process back to par.

Fig. 7-24 illustrates the point: Alternative 1, termed end-of-line quality measurement, is the typical method of measuring the quality of the finished product or service. This method usually works but issues and problems may arise often as surprises in poor product quality or service at later date. Alternative 2, termed process-tracking, is recommended. This process tracking allows us to track processes early and often gives a warning of upcoming problems; process-tracking can be a leading indicator.

A business example will illustrate this point: If housing loans are given to anybody, regardless of income or job stability, then at later date some borrowers may default on the loan. If we only measured defaulted loans, we will discover the problem when it is too late to react. Hence, measuring the quality of loans (for example via the borrower's ability to pay), will provide a leading indicator via process-tracking.

To reiterate, our belief is: *A good process will give good result, whereas a poor process will yield poor results.* Hence the best way to assure good and consistent performance is to manage, measure, and ensure good processes via internal process-tracking.

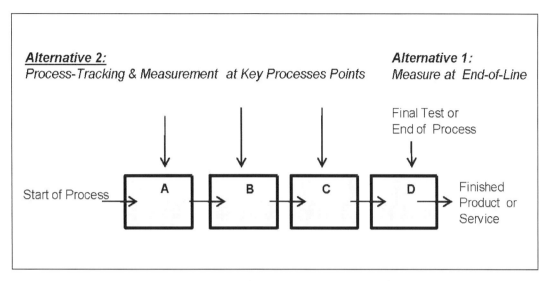

Figure 24: Alternatives for quality measurement of a process

Total Quality Management (TQM)

TQM is an effort that is committed to quality, driving improvements, and keeping customers happy. TQM focuses on providing the best possible products and services via robust processes; this will positively impact productivity, customer satisfaction and loyalty, market share, and profits. It is still very prevalent with Japanese companies. We believe its principles are essential for business success.

Objective of Total Quality Management[32]

Here's our definition:
    An effort of continuous improvement that uses quality tools and systems to manage a business, in order to make it more efficient, productive, customer focused, and more competitive.

In our context total quality goes beyond traditional product quality includes customer satisfaction and loyalty, operational efficiency and productivity, aiming for zero defect products and services, and good management of all functions in a company. *We do not propose an additional effort in Total Quality Management; we advise, however, that its principles be integrated into operational excellence.* The most important principles Total Quality Management are customers' first, quality first and total participation.

    Customers First: Customers always come first. They pay our wages and bills and help us to be successful. *"The next process is our customer"* is an extension of the customers' first concept. In an organization, everyone is part of a process and everyone has a customer. That customer must also be satisfied, not only the external customer. This creates a chain of suppliers satisfying each customer until we reach the end customer. Remember, there is no such thing as too customer sensitive but there are such things as unhappy customers or insufficient customers.

Quality First: Quality comes before anything else; there are several items of importance here:

- *Continuous Improvement.* This is a fundamental concept of total quality. We strive for continuous improvement and aim for perfection. The concept of continuous improvement is predicated on the knowledge that breakthroughs will be few and far between – hence incremental improvements must be encouraged in all processes, products, and services.
- *Attractive Quality.* We discussed this earlier. In the spirit of continuous improvement, when problems and defect levels are low, we must start providing more attractive quality items to customers. Examples of this are found in mobile phones (Apple I-phones) and the automobile industry (Toyota Prius). This attractive quality is what is defined as the second dimension of quality. *If you are routinely operating in this second dimension of quality, you will be operating in a new zone called TQC or Total Quality Creation.*
- TQC also happens when we operate at the highest hierarchy of quality of predict and prevent.
- *Aim for zero defect products and services.* Companies with low defect products and few customer complaints have a very high reputation which can improve customer satisfaction and sales.

Total participation: Employee development and participation are a very important element of a total quality effort. No total quality effort can be effective without employees contributing to improving products and processes. This includes: Quality Circles or teams, and employee suggestion schemes, and employee education. More details on quality circles/kaizen teams and suggestion schemes can be found in the chapter on employee development. Total participation also includes all departments improving their internal efficiency and productivity, and reducing costs. This is discussed further in the chapter on Hoshin Kanri planning.

Summary: Quality

In any business or organization, we need both continuous improvements and breakthroughs. With the objective of continuous quality improvement, we discussed how to drive quality improvements throughout the organization. We discussed the Kano model and two types of quality: Attractive and must be quality.

Continuous quality improvement must be done systematically by using the PDCA cycle, which is a data driven, scientific method. Within the PDCA cycle we can use the numerous statistical tools such as the seven tools and seven new tools, and design of experiments. We also discussed the use of PDCA for management reports via the A3/A4 format.

However, improvements must be maintained and for that we need to monitor and control quality on the production line via SPC, internal and supplier audits, poka-yoke, and jidoka. To achieve high customer satisfaction we need to ensure we implement the concepts of Total Quality Management, methods that help us aim for zero defects, and methodologies that drive us towards Total Quality Creation.

Often, there is a concern on investing too much in continuous improvement - because one feels that one is good enough, or already competitive, or that there is balance between cost and quality. This complacency is dangerous in any organization, because complacency is the very enemy of quality and customer satisfaction. This is expressed as the *"Whispering of Satan"* by Takoshi Hokake:

"......balance between cost and quality" is such a phrase that affects us like opium. Tearing off this veil of this beautiful phrase "balance of cost and quality" − the "quality first" policy should be adopted. This is the very energy source. But I have no weapon to shut out this whispering of Satan."

We leave you with the above comment as food for thought.

Chapter 8

Continuous Flow: One-Piece Production

[Flow is] being completely involved in an activity... Time flies. Every action, movement, and thought follows inevitably from the previous one, like playing jazz. Your whole being is involved, and you're using your skills to the utmost.
Mihaly Csikszentmihalyi

Overview

The superior methodology in manufacturing is to manufacture in a continuous flow process producing one-piece at a time or in very small batches. One-piece production will maximize throughput in manufacturing, minimize work in process (WIP) inventory, run with the highest quality level, and provide a system that allows for continual improvement of the manufacturing process. So far, we have discussed techniques such as 5S, standard work, cycle time, line balancing, just-in-time production, high quality, defect prevention, setup time reduction, and machine maintenance. Continuous flow manufacturing is where all the techniques we have discussed come together to provide operational excellence in manufacturing – to achieve World-Class manufacturing.

In this chapter we will discuss continuous flow and one-piece production methodology. Specifically we will review and discuss:

- The benefits and challenges of one-piece production.
- Little's Law and continuous flow manufacturing.
- Alternative production systems.
- Design and management of linear and cellular production.
- Converting from batch to continuous flow manufacturing.
- Minimizing variability in all processes to facilitate continuous flow.

One-piece Production

One-piece production refers to the methodology of processing and moving one job between workstations. Alternatively, we could be processing and moving large batches between work stations. The alternatives are illustrated in Figs. 8-1 and 8-2.

156   *Winning with Operational Excellence*

If we manufacture picture frames, then in Fig. 8-1, we see how a batch of frames is processed through 5 workstations. If the batch size is 5 pieces and the workstation cycle time is 1 minute for all stations, then it will take 5 minutes to complete the first batch at station 1. The batch will move to station 2, and it will again take 5 minutes to complete the batch. Therefore, as the batch takes 5 minutes per station, it will take 25 minutes before the first batch of 5 frames is completed and exits the production line.

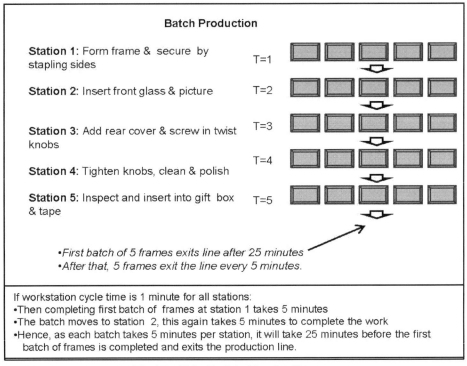

Figure 8-1: Batch Production

Why manufacture in batches? It comes from the belief of economy of scales – the larger the quantity built, the more productive the operator because of repeatability of the same routine over and over again. But is this really an efficient way to manufacture? We will answer this question.

In Fig. 8-2, we see how the same work is processed with one-piece production by building one frame at a time. Again we assume that the workstation cycle time is 1 minute for all stations. Then completing the first frame at station 1 will take 1 minute. The first frame will then move to station 2, and it will again take 1 minute to complete. Hence, as the frame takes 5 minutes to move through all 5 stations, it will take only 5 minutes before the first finished frame exits the production line. However, the 5th frame will exit the production line at the 10th minute, which means the first 5 frames will only take 10 minutes to exit the line. Compare this to the 25 minutes it takes to complete the first 5 frames, with batch manufacturing.

To summarize: the batch process takes 25 minutes for 5 frames to exit the production line whereas with one-piece production it takes 10 minutes for 5 frames to exit the production line. There is no magic here, it still takes 25 minutes of effort to complete the same 5 frames when we use one-piece production; but one-piece production is more efficient as we are not encumbered by having to work on a large batch at each station. One-piece production is applicable for all types of small or large machine assembly, work cells, and final assembly lines.

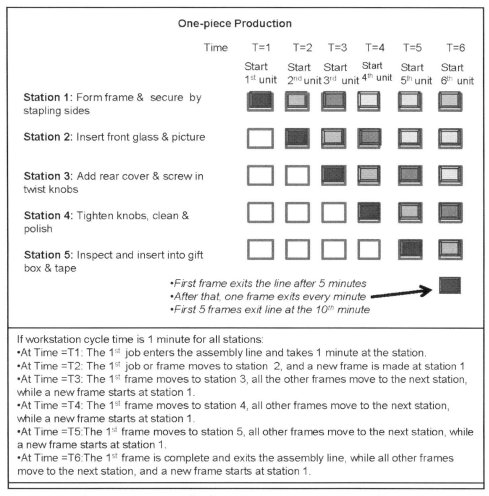

Figure 8-2: How one-piece production develops into continuous flow manufacturing

Benefits of One-piece Production

Less work in process inventory: In the batch process illustrated in Fig. 8-1, each station carries an inventory of 5 frames, while the entire production line has an inventory of 25 frames. But in the one-piece production line, each station will process one-piece at a time and hence will have an inventory of 1 frame per station and the entire production line will have an inventory of 5 frames. This is an amazing reduction in work in process (WIP). In a large factory one-piece production will permit a huge reduction in WIP and overproduction.

Reduction in factory space: With the reduction in WIP, there will be a corresponding reduction in space required to store the WIP between workstations. This reduction in space will be further enhanced with a *kanban* material system.

Faster response to changing customer demand: One-piece production allows for speedier production flow and reduction of assembly lead-time. In Fig. 8-2, we illustrated how it's possible to move product faster through the production process via one-piece production. In a factory manufacturing complex products this can result in quicker delivery to the customer.

Increased productivity: When we implement one-piece production, we can monitor and manage each work station. Using the takt time system and standard work, we can balance the production process in a fast moving linear assembly line or a slower cell assembly process. As a result we will have a more productive production process and less wasted time at each operation, with each operator working at maximum efficiency.

Higher quality: There are several ways by which one-piece production will improve quality. First, one-piece production will result in quick detection of errors and defects. Refer to Fig. 8-1: When a defect is found at test or inspection, the entire batch that exits the production line could be bad and will be rejected and require rework. There could also be lots of WIP upstream that is defective. However, in one-piece production, when a defective is found only that unit could be bad. As WIP is low, the impact upstream on the production line would be minimal and limited to a few pieces. At that point corrective action can be taken to go upstream to locate the station where the error occurred and fix the root cause of the defect quickly. Second, the philosophy of one-piece production and continuous flow requires an effective *jidoka* system, which can prevent the production of defective products and directs attention to quickly resolving the problem.

Continuous flow manufacturing: When one-piece production is running smoothly, we will have a continuous flow of work-pieces down the line. This is illustrated in figure 8-2. Continuous flow has been used to describe an industrial process such as oil refining where raw crude oil goes in one end and various grades of gasoline exit the other end. In today's context, continuous flow also refers to one-piece production which creates a continuous flow of work in a factory, with raw material entering at one end and the finished product exiting at the other end.

Challenges with One-piece Production

One-piece production requires effort and hence there are many challenges and roadblocks, which are real but surmountable.

There will be frequent disruptions and line stops whenever quality is bad: In batch production mode, if rejects are found exiting the line, we simply sort out the bad and move the good work-pieces along to the next stage of production. Work continues as usual and the defective work may pile up but will be sorted and resolved when time permits. In one-piece production, if rejects are found exiting the line at inspection or test, we prefer to stop the line and look for the cause of the defect. Typically, there is a trigger point of (say) 3 defects when a line must be stopped. This process of analyzing problems and identifying the root cause can be disruptive and if employees are not trained to deal with such a process it can cause confusion and line stops. Hence, oftentimes the start of a one-piece production process ends up in frustration or failure.

Unbalanced cycle time will create bottlenecks along the line: When a bottleneck occurs, it will cause delays and frustration at points where cycle time is too high and bored operators where cycle time is too low; whereas for a batch process everybody is busy building lots of WIP. Good implementation of one-piece production requires an effective takt time system and accurate work standards with cycle times within 10% of each other.

Material shortages will cause line stops: Effective implementation of one-piece production will require less WIP and fewer raw materials on the line. Hence material shortages due to delivery delays and problems will cause line stoppage. Therefore, when one-piece production is implemented we must introduce an effective kanban delivery process. Even if this issue is difficult to resolve initially, we can still implement one-piece production but as problems occur we must fix them quickly.

Line stops and disruptions can occur if a trained operator at a workstation is absent or missing: This can be fixed by ensuring operators are cross-trained and able to multi-task. Floating senior-operators, who are multi-skilled, are often deployed for such a contingency.

Lengthy changeover times can disrupt one-piece continuous flow: In high volume production where a line runs for days with one product, the time taken to changeover to manufacturing a different product may be less of an issue. But changeover becomes an issue during low volume production as the time taken may be very long relative to time to build the small volume of production. Changeover time delays can be due to:

- Machine or equipment that is not designed for one-piece flow. This can happen if some equipment only runs batches: For example heat ovens to age, temper, or test products. The solution here is to minimize batch sizes, find ways to reduce time taken in the batch process, and add parallel equipment to speed flow, or add buffer inventories to ensure subsequent stations are not starved of product.

- Machines and equipment which take a long time to setup: for example jigs, dies, presses, plastic molding equipment, or SMT equipment. The solution is to decrease set-up times.

Little's Law: Reducing Queues In the Factory

John Little[33], a professor at MIT, proved a mathematical theory on the behavior of queues in a system. He stated: *The long-term average number of customers (waiting in a stable system) is equal to the long-term average arrival rate multiplied by the long-term average time a customer spends in the system.* In other words:

$L = \lambda \times W$ *(arrival rate x waiting time)*

Where
L = number of customers in the system
λ = long term arrival rate of customers in the system
W = time the customer spends in the system (say a department store or bank)

Little's law applies to any system or subsystem within a system. The law has applications in queues in software design, customer service lines, logistics, and factory assembly lines. The law can also be applied to queues in a factory according to Hopp and Sherman[34]. Therefore:

$WIP = TH \times CT$ *(or Throughput × Cycle Time)*

Where:

WIP = Work in Process. This is the average number of items currently in production. This is the quantity in the assembly line from production start to finish.

TH = Throughput. This is the average output of production. This is calculated by determining how many items are produced and dividing this by time taken to produce them.

CT = Cycle Time. This is the time taken to complete the production cycle, from production start to finish.

Little's law applies to an entire factory, an assembly line, or an individual workstation. Hence when computing data it is important to ensure the unit of measurements are correctly stated. With this law several beliefs that we have taken for granted can be proven from the equations:

$WIP = TH \times CT$ .... Eq. 1

$TH = \dfrac{WIP}{CT}$ .... Eq. 2

$CT = \dfrac{WIP}{TH}$ .... Eq. 3

- According to Equation 1, if we wish to reduce WIP, we must reduce cycle time.

- According to Equation 2, if we wish to increase throughput and manufacture more product, we need to reduce cycle time. Based on our experience, this is obvious.

- According to Equation 3, if we want to reduce overall factory cycle time, we must reduce WIP in the assembly line. But more important, we must also reduce total WIP in the entire factory or system. The WIP can be in many forms: Waiting in queue for processing, in transport, defective products, awaiting machine setup, or waiting for a machine to be repaired. To reduce WIP in the system, we must: Implement Kanban, reduce setup time, eliminate wasted time, ensure that machines and equipment are running with no losses or stoppages, and quickly detect and fix production defects. As a result, for the same throughput, we can lower both inventory and cycle times with proper production strategies.

- Equation 3 also says that if we improve output at a workstation (which has an unlimited supply of material) and do not limit production, then the extra parts produced will clog a downstream station and increase overall cycle time.

Little's Law and Continuous Flow Manufacturing

We illustrated the impact of one-piece production in Fig 8-2. As we stated it's not magic; in fact it is supported by theory, specifically by Little's law. See the adjacent box for a detailed discussion on Little's Law.

To increase throughput, that is manufacture more product, we need to decrease cycle time: this is evident from our own work experience and supported by Little's law. We start by reducing cycle time at each workstation.

However, overall cycle time in the factory can also be improved by looking at other processes in the factory. Little's equation makes it clear that if we wish to reduce overall cycle time in the factory or assembly line we must reduce WIP. This specifically relates to the total WIP in the factory, assembly line, or workstation. Hence we need to look at all available options to reduce WIP using all of the techniques we have discussed so far:

- Reduce unnecessary inventory on the factory floor: Deliver production parts to assembly via a Kanban system.

- Reduce inventory piling up due to equipment issues: Ensure equipment and machines are running with no losses, slowdowns, stoppages, downtime, or poor maintenance.

- Reduce inventory awaiting machine startup: Ensure short setup time at machine changeover.

- Reduce inventory due to model changeover: Ensure kanban system and operators respond quickly to model changeover.

- Reduce defective products: Ensure quick detection and resolution of defects.

- Eliminate all forms of wasted time including motion and walking with improved standard work.

- We can also reduce wasted effort and time by improving production layout in the factory. Let's look at some alternative layouts.

Alternative Production Systems and Layouts

The objective of a good production system is to minimize transportation loss, lower costs, and increase productivity. In chapter one, we discussed the seven wastes including transportation loss: In a factory we should move material only when required with a minimum of effort. Note, however, that we are not promoting automated conveyors and robotic transporters; instead we should always look at the most cost-effective method to minimize transportation loss and cost. Good and well planned factory systems are the best means to achieve this objective. Furthermore, an optimum factory production system can increase productivity.

162   *Winning with Operational Excellence*

	Product Type			Competitive Output Parameters				
Type of Production System	Few Products, High Mix	Low Volume, High Mix	High Volume, Low Mix	Delivery	Cost	Quality	Flexibility	Innova-tiveness
1 By Equipment Function	Job-Shop	Batch		Acceptable	Poor	Poor	Good	Good
2 Linear or Cellular Flow		Small Batch	Small Batch	Acceptable	Poor	Poor	Good	Good
3 Cellular, Paced by Cycle Time		One-piece continuous flow		Good	Good	Good	Good	Good
4 Linear or Linear U-Shape, Paced by Cycle Time			One-piece continuous flow	Good	**Best**	Good	Poor	Poor

Figure 8-3: Product and Process Matrix, with potential benefits (Modified from Miltenburg)

A useful classification for production systems for consumer and industrial products is the product-process matrix, proposed by Miltenburg[35]. This is shown in Fig. 8-3. The figure shows several popular factory production systems. Some comments and comparisons are in order:

1. The first type is the *equipment-based layout*, which can be laid out in cells or groups. This is historically the oldest production methodology but still popular in a job-shop or small batch production; this can range from one job or small volume production each month. A production system that is *equipment-based* refers to a layout that is planned by the equipment (or process) required for manufacturing. For example: For manufacturing of metal jigs and fixtures, we could layout by lathes, drill-press, grinders, milling machines, sheet metal fabrication, and common storage area. For manufacturing of large home-appliances, we could layout by various sub-assembly stations, test centers, heat treatment stations, spray-paint stations, final assembly stations, quality inspection, and packaging stations.

   - In this equipment layout mode, work-pieces move in batches according to the equipment required for processing; there will be WIP at each station. Such a process is not optimum; hence, in Fig. 8-3, item 1, under the heading competitive output parameters, we see that this layout gives acceptable delivery, but poor cost and quality. However, it can provide good (high) flexibility and innovativeness. These parameters are relative to the other types of layouts shown in the figure. The alternative to *equipment-based* layouts is *product-based* layouts, discussed in the next three layouts.

2. Next there is the *linear or cellular flow in a batch mode*. Here the product flows in batches from station to station, and the production process is *product-based* which refers to a layout that is planned for the specific production process: Here the layout is by the actual sequence of production steps, where the work-piece moves and develops into the final product. We showed this in Fig. 8-1. In this mode, the stations are disconnected from each other and there is WIP, at each station and between stations.

    - Because production is in a batch mode and the stations are not balanced there will be a bottleneck at the station with the highest cycle time; the best option is to keep the stations in near-balance to maximize efficiency. This mode is very common and used for manufacturing heavy equipment and bulky home-appliances. In this mode, the competitive parameters of cost and quality are poor but flexibility is high. Refer to Fig. 8-3, item 2. A point to note: When an operation moves from an equipment based layout to a product based layout, the amount of equipment that is required can increase due to duplication on more than one line or cell; hence the overall machine utilization may be reduced. This is waste, but the use of multi-skilled operators and increased efficiency with this layout will help compensate for the higher cost.

3. A very effective layout is the cellular flow, paced by cycle time. Such cells are product based, and will have dedicated assembly equipment. Here the layout is by the actual sequence of production steps, where the job moves continuously and develops into the final product. In this layout, the material flow is visible at a glance. Cellular layouts have small footprints but may consist of numerous cells; they can be arranged to minimize operator travel and allow for easy multi-tasking by a single operator.

    - A cellular layout, paced by cycle time, using one-piece production is ideal for products with small volumes and high mix or high product options. This mode works for industrial or consumer products, and is often used to build and supply parts to a final assembly line building automobiles or consumer products. This type of layout can give very competitive output parameters for delivery, cost, quality, and flexibility. Refer to Fig. 8-3, item 3. Cells are an excellent choice for high mix low volume production.

4. Another effective layout is the linear flow, paced by cycle time. In this *product-based* layout, the product moves down the line according to the sequence of production steps, where each job moves continuously and develops into the final product. Such lines can also minimize transport and material flow; assembly starts at one-end and the finished product exits the other end. Because such lines are fixed, often with conveyors, they are not ideal for frequent demand changes or changing workloads. Such lines are inflexible but suitable for high volume and low mix (few options). This mode

of production is typical in final assembly of automobiles or consumer products. This type of layout can give very competitive output parameters for delivery, and quality, with the best cost; refer to Fig. 8-3, item 4. An alternative to linear is the linear U-shaped line when more stations are required or there is space constraint. Some additional comments:

- Linear U-shaped line layouts are better than linear lines as they are not fixed by the length of a building and adding as many linear U-shaped lines as necessary is possible for complex product manufacturing.
- When assembled products are small to medium sized, both linear and U-shaped lines can reduce cost and improve flexibility by shedding bulky and expensive conveyors. Instead we can use tables that lock in like Lego blocks and pallets with casters.

In an efficiently managed factory, final assembly will be run in linear lines or multiple cells. Operators at each station will have the material, components, and sub-assemblies at hand to manufacture products in one-piece assembly enabling a continuous flow of products.

The final assembly production will be fed via a kanban system. Sub-assemblies or parts may come from other assembly lines, cells, or external suppliers. Often, due to equipment limitations, sub-assemblies may be manufactured via batch processes. However, the recommended method is always one-piece production at all stages of manufacturing.

Each of the layouts discussed has its benefits. Because of different demands of customers, product mix, and options, most factories have a combination of the layouts discussed here. The challenge facing an operation is to do all of these well and still provide efficiency and quality.

Linear and Cellular Production

Choosing between Batch, One-piece Cell, or Linear Production

There are several decision rules available to help choose whether a product should be manufactured in a job shop, batch, or one-piece mode. We quote one decision rule out of the many available in the literature as a general guideline. According to Askin and Standridge[36], if n is the number of products and P is the hourly volume:

- If $1 \leq n \leq 5$, and $1 \leq P \leq 1000$, then use a paced linear line
- Else if $5 \leq n \leq 100$, and $1 \leq P \leq 50$, then use a paced cellular process or batch process
- Else if $100 \leq n \leq$ several thousand, $P \leq 1$, then use a job shop or batch process

These are broad guidelines and starting points in the decision process; the guidelines provide a direction to work towards. The critical parameter is *n*: the number of products built.

Basically, if n is low and P (hourly output) is high, the direction is clear: use a paced linear line. But if n is large or if P is low, we would take the top 10 products in terms of volume and start to convert them to one-piece cell production. The experience gained will set the stage for other improvements. The goal should always be to move towards one-piece production.

Design of Linear Production Line

In the chapter on cycle time, we discussed a line balancing project. Such lines are typical for high volume production, where one-piece production is the norm. These are typically high volume lines, delivering 30 or more products per hour, with takt times of three minutes or shorter. The line can also be designed to allow for mixed model production and not just high volume production with just one or two models. In fact the Toyota Production System (TPS) was developed to manage mixed model production and not high volume assembly lines.

Fig. 8-4 shows a paced linear production line. The line produces industrial printers working at a takt time of about 120 secs. The assembly process is fairly complex and requires strong skills. The product has many options, often requiring 10 to 20 configuration changes daily. This requires good standard work and multi-skilled operators. Note: Parts required for assemblies, such as screws, gears, and springs are stocked in front of the operators and delivered via a kanban process. Sub-assemblies for the final product are delivered via kanban, and can be seen in the center of the line. The work piece moves from left to right on a pallet with wheels. This overall layout minimizes operator travel and motion.

Paced linear line: Once the decision has been made to select a paced linear line, you need to design the process to convert from batch, prototype, or NPI (new product introduction) lines to paced linear production. The following are guidelines:

1. Typically you start planning for the final assembly line, as that will set the stage for the rest of the operation. Final assembly will pull from earlier sub-assembly stations, cells, or suppliers.
2. Have you selected the right products per the decision rules given in the previous section? Confirm the production volume and determine takt time for the line.
3. Determine the workstation and number of operators. This is driven by the takt time and total cycle time. If the number of stations/operators is too many for your final assembly line, reduce the total cycle time (for final assembly) by moving some production to sub-assembly lines or cells.

4. You need to decide on whether to setup one or more linear lines. The solutions include multiple lines or U-shaped lines when takt time requires too long an overall cycle time, and hence a very long production line.

5. Once the production steps are confirmed, prepare and review standard work for each station. When designing the standard work, take into account operator work time, handling time, movement and walk time. A work combination chart for complex stations will be useful – refer to the chapter on standard time. Are the cycle times for each station within 10% of each other? This information can be obtained during the prototype design and run of a new product. Remember that longer cycle times can lead to boredom and errors; this can be minimized by regular job rotation done daily or every few hours.

6. Determine the equipment required. Can your equipment at workstations meet the required cycle times? Do you need parallel equipment? This can be the case for a station which cannot be split into multiple stations, for example test or complex assembly procedures. This may necessitate multiple parallel stations.

7. How many operators are needed to meet takt time? How will you train and distribute the work among operators – between complex Vs simpler assembly stations? How many operators need to be multi-skilled?

8. Review material delivery and kanban requirements. Use card or signal systems for material replenishment.

9. Determine Test Stations and Quality Assurance points. Review expected or current yields and set targets to be achieved.

10. Prepare a process flow and layout taking into account all of the above points, including station cycle times plus kanban and inventory locations.

11. During the pilot run, work towards balancing the workstations. This may take several iterations. The engineering and production team need to review and detect glitches and difficulties, and ensure quick intervention and resolution.

12. If you are planning *mixed- models production* (run product A followed by product B), note:
    a. Mixed-model production is feasible if work content is similar or workstations use similar equipment with different standard work.
    b. Test stations or equipment can be setup quickly for different models.
    c. Ensure parts for different models and tools are provided if required
    d. As discussed in the chapter on standard work, a good procedure for mixed model production is to announce model changes via a moving model change box. This is also a very effective method of completely removing WIP of the current model and starting quickly with a new model.

Continuous Flow: One-Piece Production   167

Figure 8-4: Paced linear production line

Cellular Production

The motivation behind paced work cell system design is to implement continuous flow with small mixed model volumes, with low WIP, high quality, and efficiency. The many characteristics and benefits of cells include:

- A work cell is a work unit larger than one workstation or a piece of equipment; the cell layout or footprint can be small to large. Typically, it can have from two to 15 operators and as many workstations. Multiple cell operations can be combined to form an overall cell production farm, which can be optimized for flow.
- Flexibility: Ability to change number of workers as production volume changes. This requires that operators are trained in multiple skills and be familiar with various processes and equipment.
- Cells are ideal for multiple and complex equipment layout, as each worker is required to multi-task several machines at one time, with minimum of walking. This raises productivity.

- Cells can be structured to handle multiple products, that are similar, or products requiring similar workstations or machines.
- Cells are superior to linear layouts when multiple and complex equipment is required in production. This is because cells are more flexible for changing production volumes and WIP inventory control.
- Typically the entrance and exit of a U-shaped cell are close together; if necessary WIP can be controlled with a new unit starting production (by entering the cell) only when one unit leaves the cell. Within the cell work pieces can move from one station to the next, without intermediate storage. Cells can operate in small batch flow or one-piece flow. However one-piece flow is the recommended strategy unless the cell operates with equipment that requires small batch production, such as casting or heat-treatment equipment. When batch equipment is used it is best to run the smallest possible batch via short setup times.
- Cells are able to handle frequent model changes, but the setup time for equipment change is critical. For zero to low equipment setup times, model changes are easy to manage. When setup is high, the minimum batch size should be produced before switching to another model.
- The important characteristic of the cell is that it has a cluster of workstations that maximizes operator communication that minimizes operator travel.

Major Benefit of Cellular Production

A major benefit of cells is *enhanced teamwork*. The close proximity of operators allows for good communication; furthermore operators will be able to assist each other when needed. When there are bottlenecks or high WIP the feedback is immediate and operators can work with engineers and supervisors to improve the process. This type of camaraderie is seldom possible in linear production lines; hence we have seen that cells are often more efficient and productive than linear lines.

Design of Cellular Production

Once the decision has been made to select a cell system, you need to design the process or convert from equipment or batch based to product and cell based production. The following are guidelines to get started:

1. Confirm the production volume and determine takt time for the cell. You could be planning for the final assembly cell or for a cell to support a final assembly line.
2. Decide on the type of cell flow required; baton, line, or chaku-chaku. This is discussed in the next section and is dependent on the product type, the equipment in the cell, and the product demand.

3. Determine if you require multiple stations for complex operations and equipment.
4. Determine the production steps and processes needed. Determine the workstation cycle times.
5. Once the production steps are confirmed, prepare and review standard work for each station. This information can be obtained from an earlier prototype production stage. When designing the standard work, take into account operator work time, handling time, movement and walk time. A work combination chart for complex stations will be useful – refer to the chapter on standard time.
6. Determine the equipment needed and the setup time for any equipment that requires batch production. This will influence how you synchronize this station with the rest of the cell and if you need to have small inventory buffers.
7. Determine how many operators are needed to meet takt time? How will you train and distribute the work among operators. How many operators need to be multi-skilled?
8. Review material delivery and kanban requirements. Use card or signal systems for material replenishment.
9. Determine Test Stations and Quality Assurance points. Review expected or current yields and set targets to be achieved.
10. Prepare a process flow and layout taking into account all of the above points, including station cycle times plus kanban and inventory locations.
11. During the pilot run, work towards balancing the workstations and the operator workloads if they are multi-tasking. This may take several iterations, and will require continuous review by the engineering and production team to detect glitches and difficulties, with quick intervention and resolution. Note:
    a. For mixed-model production, managing the product sequence is very important. The challenge is different cycle times and the impact on line balancing and subsequent cell or assembly lines.
    b. The comments on mixed-model production mentioned earlier also apply here: Manage equipment setup times and ensure sufficient parts and tools for different model assembly. The heijunka scheduling process can be applied successfully for mixed-model production.

Cellular Production Layouts

Fig 8-5 shows a U-Shaped Cell. The operator at the entrance is able to control what goes in and out, hence ensuring control and management of WIP. The number of operators needed depends on the volume, cycle time, and process requirements.

170   *Winning with Operational Excellence*

- In Fig 8-5a there are 6 workstations, with 6 operators; we call this a fixed-station, or baton-touch, mode as product is passed from operator to operator.
- In Fig 8-5b market demand for product has dropped by 50% and we need only half the number of operators, hence each operator doubles up and works on two stations each.
- Fig. 8-5b is typical of what needs is done when there is low volume production with complex assemblies or equipment. Here the operator will move with the job from station to station. The operator performs machine setup, loads a job, completes the job then moves the completed job to the next station; once again the operator does machine setup, loads the job, and then returns to the previous station to repeat the cycle. We call this a line flow mode.

Cell layouts can range from U, C, L, S shapes, and parallel configuration. A cell can be a simple short line or a back to back short line. Figure 8-6 shows a simple U-shaped cell. Complex and larger products may require more innovative configurations, as in Figs. 8-7 and 8-8.

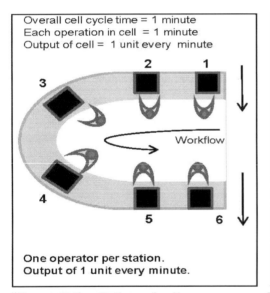

Figure 8-5a: Cell production

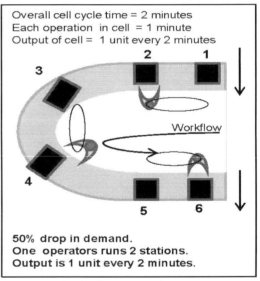

Figure 8-5b: Cell production with lower demand and multi-tasking

Continuous Flow: One-Piece Production   171

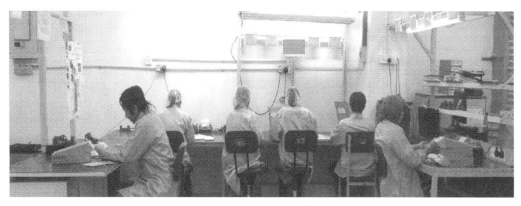

Fig. 8-6: Simple U-Shaped cell.

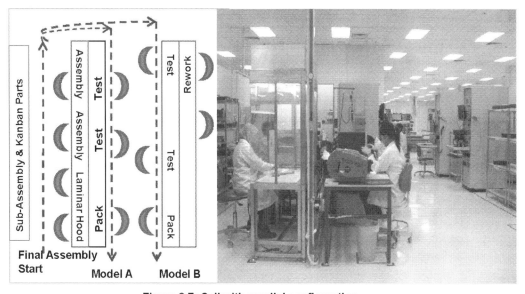

Figure 8-7: Cell with parallel configuration.
The flow-chart on left shows the layout.   Currently, only Model A is in production.

Figure 8-7 shows a production cell with parallel configuration. In this case, sub-assemblies arrive at the cell and are assembled into two alternate models – model A or model B – based on customer demand. The product is a RF Analyzer and the finished model undergoes extensive testing and calibration after which it is packed for shipment. As the test equipment is unique, there is dedicated test equipment for each model. The production operators are multi skilled and able to work on both models.

Figure 8-8 shows a large U-shape cell which handles heavy assemblies. The sub-assemblies enter the cell on the left side and move from station to station on trolleys. The finished product exits on the right side. Note the 5S system: neat layout, good overhead signage and floor with ESD and location markings.

Figure 8-8: Large U shaped cell, handling heavy assemblies. Sub-assemblies arrive on the left side and move from station to station on trolleys. Units in the center are awaiting final QA, while button-up and ship out is on the right. Note: Inserted picture on the left shows operation in the first station on the left after entering the cell.

Types of Cell Flows

From our experience there are several types of cell operations. The specific type will depend on the assembly process.

Baton touch or fixed station: Here each workstation is independent and the operator will complete his or her operation and then pass the job to the next operator. This is typical for small volume assembly operations or where the volume demand is high enough to keep all stations busy. This is similar to Figs. 8-5a and 8-7; it's also similar to the assembly process on a linear paced line.

Line flow: Here the operator will move with the job from one station to the next. The operator performs machine setup, loads a job, completes the job then moves the completed job to the next station; at the next station the operator again does machine setup, loads the job, and completes the job. She can continue going down the line or pass the work-piece to the next operator as in Fig. 8-5b, and then return to the previous station to repeat the process. At each workstation she will perform the required task and then move the work-piece to the next station. There could be one operator in the cell performing all tasks for low demand. Alternatively with higher demand there could be several operators moving in line flow. Such a process can reduce boredom of operators but requires multi-skilled operators.

Chaku-Chaku: This is translated from the Japanese as "load load". This type of procedure is typical in a cell which has processing equipment at all stations in the cell. Here one operator moves through the cell loading a machine at each workstation. This is done in sequence. The unloading may be done by the machine or by the operator when she comes around the next time. The process repeats as the work piece moves to the next station in the cell. Refer to the appendices for more details.

Maximizing Workflow Productivity

The engineer who designs a linear or cell production system often has to determine if the line should flow clockwise (CW) or counter clockwise (CCW) and if the operator should be in a sitting or standing position.

Is there are difference in productivity? Yes – according to many experts. It seems that the practice of counterclockwise flow was originally initiated by Toyota Motors in their machine shops. By standardizing the direction of material feed they ensured that material could be passed efficiently between machines. A counterclockwise direction was selected because the majority of people are right handed and moving from left to right seems natural.

Research shows that an individual's right brain processes spatial recognition and therefore human perception of space is stronger through the left side of vision (the right brain controls the opposite side of the body). When an athlete runs "left hand inside" or counterclockwise, he has better visibility of space on the left side and he is able to run more comfortably, confidently, and quickly. This is supported by the fact that all running tracks run CCW; the CCW track-layout was standardized after experiments in both directions.

Another consideration is whether operators should be sitting or standing while working at their respective stations. In both cells and linear lines it is recommended that operators work in a standing position. The standing position allows operators to be more flexible and productive: he or she can easily reach for parts or move assemblies while standing. Furthermore, a multi-tasking operator who has to move between several workstations should definitely be in a standing position. Typically a standing operator can be 10-15% more productive than in a sitting position.

Converting from Batch to Continuous Flow Manufacturing

The choice between one-piece cell and linear production depends on volumes and equipment complexity. If you have high volumes with low mix, typical for consumer products, linear or linear u-shape lines are best; such lines can also handle mixed models if the product conversion process is quick. Products that require multiple and complex equipment, or run at low volumes with high mix, are best for cell production.

Once you have decided on converting from batch to cell or linear production, the next step is to understand how the current process is running. For a small operation, you can draw a *process map*. But if you are looking at an entire operation, then a value stream map is more appropriate. There is more information on value stream mapping in Chapter 11.

Project: Converting from Batch to Continuous Flow

Here we illustrate a case of a production line that ran in batch mode. The original line started by making prototypes of a small instrument. However, this line quickly grew into volume production and improvement was required to convert the final assembly line to continuous flow. The production process started by loading components on to a printed circuit board; this was done at a SMT machine producing a batch of weekly demand. The final assembly line was also running in batch mode ever since it started original prototype production. As sales demand increased to about 300 units a day, several issues started to occur, such as congestion, delays, and increased defects.

Fig. 8-9 illustrates the *before and after process maps* of the old and improved production lines. Initially printed circuit boards were moved in large batches from the SMT machines to a different floor of the factory. Furthermore, stock of printed circuit boards was kept in batches as they were processed at each workstation. The workstations consisted of benches, where work moved from station to station in boxes.

The improved line's process is shown on the right of Fig. 8-9. The machine changeover time was reduced to 30 minutes and was run to fulfill the daily need for final assembly. The new line is a paced linear continuous flow line, where work is done one-piece at a time. Several stocking points have been eliminated. Additionally, excessive transportation and motion has been eliminated. The impact of these changes has been impressive, and is shown in Fig. 8-10. The results are typical of improvements experienced on a paced linear continuous flow line.

Minimizing Variability and Other Barriers to Continuous Flow

Variability is always present in a manufacturing environment. It will happen in individual parts coming out of the same mold all the way to the finished product coming out of the production line. To ensure smooth continuous flow, it is important that there is a minimum of variability in all operations in a factory. Any variability in the operation will worsen performance and create bottlenecks. When variability is high, the factory or production line will choke-up and throttle production.

What are some of the variability and barriers, and how do we prevent them? Here is a list from our previous discussions and some new items:

Continuous Flow: One-Piece Production 175

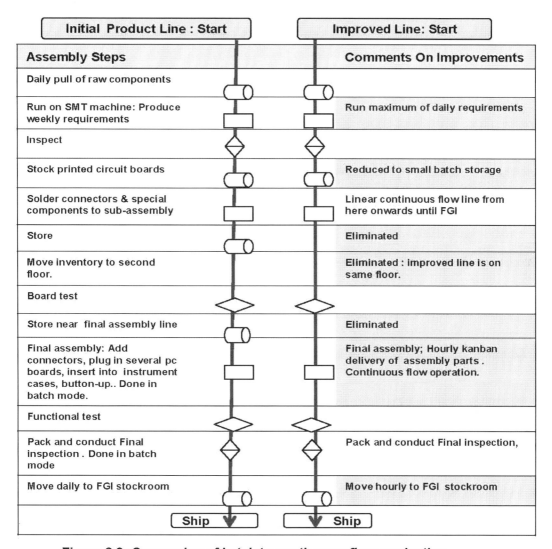

Figure 8-9: Conversion of batch to continuous flow production

Item	Original Line	New Line	Impact
Final assembly layout:	Batch on workbenches	Paced continuous flow	•Eliminated batch storage •Reduced WIP by 60% •Reduced work area by 10%
Quality	•Handling defects=2% •SMT defects=0.1%	•Handling defects=0 •SMT defects=0.01%	Big improvements due to less handling and improved machine maintenance and calibration.
Operator Productivity		Eliminated operators stocking and counting WIP + less rework	Approximately 20% reduction in operators

Figure 8-10: Impact of conversion of batch to continuous flow production

Quality and reliability of material and supplier delivery: Doing this well requires strong supplier management and an efficient kanban system. Incoming quality control of parts is of extreme importance.

Machine or equipment breakdown or downtime: Machines need to be regularly calibrated and maintained to avoid breakdowns; maintenance must be done without impacting production.

Machine setup time must be low to minimize delays in model changeovers. Low setup time will allow for one-piece production or small batch production with small buffers. Hence it is imperative that you constantly review and reduce machine setup times.

Machines with low capacity: Machines with insufficient capacity to meet daily demand will constrain the entire production flow. This can be resolved with overtime, increasing capacity, or more equipment.

Poor operator skills or absenteeism must be compensated with job rotation and multi-skill training. Multi-skill training is essential to manage quick model changes and production volume swings. Job rotation and continual skill-training will help to prevent operator boredom and encourage acceptance of the challenges of one-piece flow.

Inadequate training will result in workers doing things differently: Therefore they may introduce variation in their work, which may cause defects. Rigorous and continuous training in standard work is essential.

Defects can cause delays, bottlenecks, rework, and higher inventory on the production line. This can be minimized by aiming for zero defects and continuous improvement. Furthermore, if variation is minimized, quality gets better and inspection can be eliminated; then productivity will increase.

Improper line balancing can cause bottlenecks, delays, and inventory buildup. When a production line is well balanced, there will be a minimum of inventory and maximum productivity. But if the line is not balanced, inventory will pile up along the line and reduce throughput. Therefore, continuous review and improvement of cycle time and line balancing is a must.

Minimize all categories of WIP: Going back to Little's law, we must eliminate excess inventory everywhere in the factory to speed up manufacturing throughput.

Minimize product options: The more options you have, the more model changeovers you will need, which cause changeover delays and higher inventories. The solutions are to minimize and standardize product options. For example most consumer products now come with dual voltage 110V and 220 V.

Continuous flow on non-synchronized lines: Continuous flow is more difficult to manage on non-synchronized lines: Examples are lines which are not paced via takt-time, such as between final assembly and sub-assembly lines or a machine-intensive production floor, which wants to maximize output of each machine. In such cases the situation can be improved via the heijunka procedure we discussed earlier.

What this tells us is that minimizing variability in production will improve factory performance and efficiency. It's advisable to start at the beginning of the production process: *This is because variability early in the production process propagates downstream and gets magnified.* The concept is similar to the 10X rule we discussed earlier. Therefore first review and resolve incoming material quality. Next, review and reduce defect levels in machines and assemblies early in the production process. This ensures that you do not send defective material downstream. Finally check for bottlenecks or constraints along the line; for example look at low capacity machines that have to work overtime and poor line balancing.

Reducing variability actually helps to improve the factory output, and is therefore the first option you should look at when there is a need to increase capacity to satisfy increased demand. It is also a cheaper and quicker option that will also improve quality of the product. Hopp and Spearman[37], based on such evidence have proposed the **Law of Variability** which states:

Increasing variability always degrades the performance of a production system.

Summary: Continuous Flow and One-Piece Production

We have discussed one-piece production and continuous flow manufacturing. One-piece production is applicable for machining parts, all types of small or large machine assembly, sub-assembly manufacturing, and final product assembly lines. One piece production will maximize throughput in production, minimize work in process (WIP) inventory, help reduce factory space, provide the highest quality, and establish a system that allows for continual improvement of the manufacturing process.

We reviewed alternative production systems and focused our discussion on the design and benefits of linear paced flow and cellular paced flow. Paced linear flow is ideal for high volume production and cellular flow is best for low volume production. Both systems are able to handle mixed-model production.

However, to ensure smooth continuous flow, it is important that there is a minimum of variability in all operations in the factory because variability will worsen performance and create bottlenecks. Reducing variability actually helps to improve the factory output, and is therefore the first option you should look at when there is a need to increase capacity to satisfy increased demand. It is also a cheaper and quicker option that will also improve the quality of the product.

One-piece continuous flow manufacturing will provide faster response to changing customer demand; the manufacturing operation will also be able to respond quickly to customer demand including accepting smaller orders, which can open new market segments.

All these benefits will result in an operation that is more efficient, cost effective, and customer oriented. Continuous flow manufacturing is where all the techniques we have discussed come together to provide operational excellence to achieve World-Class manufacturing. However, these techniques can only work well if we have an enthusiastic, engaged, and educated workforce; that is the topic of the next chapter.

Chapter 9

Employee & Partner Participation and Development

Teamwork is the ability to work together toward a common vision. The ability to direct individual accomplishments toward organizational objectives. It is the fuel that allows common people to attain uncommon results.
Andrew Carnegie

Overview

Every organization must harness the potential of every person it employs or works with – including suppliers. The power of synergy cannot be overemphasized. In the first chapter we discussed the '4P' model put forward by Jeffrey Liker (refer to Fig. 1-1).

The '4P' are: Process, People and Partners, Problem Solving, and Philosophy. According to Jeffrey Liker, most companies put the bulk of their energy in implementing the P or Process step, which includes elimination of waste. However, the least emphasis is placed on the People and Partners. We believe they do this because the process step is the easiest step to visualize and do physically; furthermore they can do this step for eternity and never run out of ideas. Many TPS and lean consultants and texts also tend to emphasize on process and waste and place little emphasis on the other 'Ps'.

Unfortunately, this will result in a shop floor improvement focus, and will not harness the full potential of any organization. We have already discussed processes and problem solving in great detail. Organization philosophy is a fundamental driver and part of the Hoshin Kanri process, which we will discuss in the next chapter. Here we will touch on people and partners.

In this chapter we will discuss several methods to harness the potential of every person and manufacturing partners. This is just a start, but it moves us in the right direction. Our discussion will include the following:

- Kaizen and kaizen team activity.
- Employee suggestion scheme.
- Employee education.
- Partnering and working with suppliers.

Kaizen and Kaizen Team Activity

Continuous improvement is a never ending process because we strive towards increasing customer satisfaction, reducing waste, lowering costs, and better quality. To quote Dr. Deming: *"We need never ending improvement to establish better economy"*.

Progressive companies throughout the World have adopted this philosophy. However, do note that the kaizen process encourages small and quick improvements in an operation. But to be successful a company needs both kaizen and innovative breakthroughs. We discuss breakthroughs in the next chapter on doing the right things.

Every employee needs to participate. Teams of well trained employees on the production floor can solve process and product problems; engineering and management teams can solve service, product or organization problems. Kaizen projects can be implemented by individual performers, engineers, mangers or by teams. Teams can consist of engineers or operators. A Quality Circle consisting of operators is a very powerful way of getting operators to be involved in quality and productivity improvements. Here are some guidelines based on our experience:

Management of Kaizen teams and activity: Kaizen teams and improvement activity should not be operating ad-hoc; therefore an overall structure and direction is necessary. This is not a staff or consultant activity; operation or engineering managers must steer and drive this activity. One approach is to have a continuous process improvement (CPI) team in each operation or production line. There may be several CPI teams in a large product division, with each team focusing on an operation or product group. The teams meet regularly (at least weekly) to review customer issues, shipment delays, process defects, process yields, internal productivity metrics and targets, and operational objectives; from these they must select areas requiring improvement.

Appoint a team and team-leader: Depending on the nature of the project, the team can be a small kaizen team consisting of people from the same work area: For example from operators, technicians, and engineers. This group would be led by their supervisor or an engineer. The team can also be a higher level cross-functional team consisting of professionals or managers from different work areas: For example, employees or managers that form teams from different departments, different divisions, or between a company and its suppliers.

A team facilitator may be needed to help them with the improvement process. This is necessary for a new team, while an experienced team will already have the necessary skills-set. The facilitator is an experienced and well trained individual. He or she can assist the team in staying on schedule, understanding the PDCA cycle, developing practical solutions, seeking technical help, and educating members. The role of a good facilitator includes:

- Help the team identify the right issue and project to work.
- Help the team understand the PDCA cycle and how to apply it.
- Challenge the team to select aggressive improvement targets.

- Provide overall guidance: Encourage good data collection and analysis, development of a strong hypothesis, and proposed solutions.
- Help the team develop strong decision making ability.
- Teach the team to do productive gemba walks so that they can identify waste and recognize opportunities for improvement.
- Ensure the team completes the project with solid standardization (Act stage) to prevent problem recurrence.

Provide education and training: All teams should get basic training in the PDCA cycle from an experienced manager, engineer, or the facilitator. With good progress, there will be a need to provide additional training; this can be requested by the team leader, facilitator, or team members. A list of completed projects should be available and accessible as a learning tool.

Improvements projects can be of several types: In the chapter on quality we mentioned that for production teams these projects can be driven by customer complaints and inputs, and internal quality issues: One such project was discussed during the detailed PDCA discussion. Kaizen projects can include:

Reduction of the seven wastes on the factory floor. For example: overproduction, unnecessary transportation, unnecessary motion, improving standard work, SMED projects, reducing excess inventory, reducing quality defects, and improving yields.

Small day to day improvements on the production line or warehouse; these can be quick improvements suggested by operators, technicians, or engineers. This can be a tweak of standard work or production equipment that improves productivity. Such projects may not have major productivity improvements but encourage employee engagement and participation; and create a healthy and stable workforce.

Schedule activities and monitor progress: All teams should prepare a project schedule and improvement target. Any deviations or delays will require intervention and analysis by the team-leader or facilitator. Teams must keep minutes of all meetings and track the project milestones.

Recognize and reward: Ongoing recognition and rewards are a must. An excellent and powerful way to communicate complex or simple projects is to summarize them on a Kaizen template. These projects can be displayed. Such displays and recognition convinces both customers and employees how serious the organization is about quality and productivity. Fig. 9-1 shows a Kaizen display board. In this example, each project is summarized on a one page format showing project benefits and improvements. Both simple and breakthrough projects can be displayed. Rewards can be decided by management and HR based on company policies.

An important afterword on kaizen and quality teams: When the concept of quality circles or kaizen teams was first introduced, participation was voluntary. In today's highly competitive environment, everybody must chip in. Participation is no longer voluntary – it is essential for success. Hence, kaizen teams must be encouraged at all levels in the organization; this includes formation of management or cross-functional teams.

Figure 9-1: Kaizen project display board

Employee Suggestion Scheme

A successful employee development and participation effort will require the participation of every person in the organization. We must never underestimate the worth of each individual. Consider the following news article[38]:

> *If you think you are working hard, read on:*
> *...A New Record of 9310 Suggestions per Year*
>
> Mr. Koji Nakayama, QC Section at Utsunomiya Plant of Matsushita Electric TV Division, has established record-high improvement suggestions reaching a total of 9310 suggestions in a year, all by him. When asked about the secret of his inexhaustible source of suggestions, Mr. Nakayama says: "When an improvement idea hits me, I will go over it little by little on a daily basis. I try to put my idea in a presentable form and suggest it daily. I work on the idea on an average three hours every night after returning home.
>
> Are there complaints from his wife and kids? "No problem," he says, "because I put myself at their disposal every weekend. So they don't bother me during weekdays". As to the source of his ideas, "I'm a voracious reader and read all those newspapers, magazines, journal articles and watch TV programs that have something to do with my work directly or indirectly".
>
> "Good communication with other departments in the company is important to get richer information. Another source of suggestions is to develop (new) ideas for improvement after (each) improvement. I always keep a memo pad with me to jot down my ideas and observations."

In Japan, many companies obtain an average of more than 20 suggestions per employee per year and up to 90 percent are implemented. The results can be millions of dollars of savings and better quality processes, products and services. At Toyota up to 98% of suggestions are implemented, because the employee consults with fellow employees and supervisor to fine-tune his suggestion. Suggestion scheme works in the West at companies like Siemens, Pfizer, and IBM.

The Foundation for a Suggestion Scheme is Twofold

First, it allows all employees to have a say in improving areas that they think are wrong, to allow bright ideas to be captured formally, and to recognize that the Company's management does not have all the answers. Second, it allows lower-level employees to engage actively with the company business.

Suggestions may range from very beneficial to incremental improvements. Rewards and payments to employee may be minimal; in Japan, most rewards are between 500 and 2000 yen, or under US$20. *However, the employee engagement and involvement is priceless.*

Guidelines for a Suggestion Scheme

Here are guidelines for an employee suggestion scheme, based on our experience. More information is provided in the appendices for suggestion scheme forms. A tested and successful suggestion scheme review, reward, and grading process are shown in Fig. 9-2.

Objectives and Mechanics of an Employee Suggestion Scheme: The objective of suggestion scheme is to support the participative management philosophy of the company. Therefore it allows employees to give logical and practical suggestions for improvement in their work environment.

What topics are applicable for a suggestion scheme? Suitable topics will include items that contribute to safety, productivity, quality, and customer satisfaction in the organization, specifically:

- Production materials and its flow.
- Safety-related issues.
- Design and layout of production floor.
- All systems and processes.
- Design of equipment, tool and fixtures.
- Work environment.
- Quality and design of products and services.
- Standard work and other work procedures.
- Information flow.
- Customer services and customer relationships.

What topics are unsuitable for a suggestion scheme? There are several topics that are considered unsuitable, specifically:

- Suggestions that are within the direct control of the proposer.
- Personnel policies and guidelines.

- Salary and wage administration.
- Personal grievances or conflicts.

What are the operating rules of a scheme? The rules of operating the scheme must be clear and allow effective implementation. Some guidelines:

- Suggestions should be submitted on a simple and standard form.
- All concerns requiring improvement in production or the office space must be accompanied by a suggestion before they can qualify for assessment. The proposer should approach her immediate supervisor for assistance wherever necessary.
- The allowable topics have been discussed above and exclude personnel issues.
- The company reserves the right to make changes to the suggestion scheme and its reward system whenever necessary.
- The judges' decision in grading the suggestion is final.

The suggestion scheme grading process: There are many ways to do this, but here is one successful way: All individual and group suggestions will be graded into thank-you, bronze, silver, or gold awards, based on a points system. The points given will depend on the following criteria.

- Idea
- Effort
- Customer satisfaction
- Net savings
- Safety
- Quality

Reward System: Remember, the purpose of the scheme is to allow professionals and lower-level employees to engage actively with the company business. The process and cost of reviewing each suggestion and giving a small reward seems costly if the transaction cost is computed. *But the employee engagement and involvement is priceless.* Hence you must devise a reward scheme with this in mind – it's not the money, it's the recognition and engagement that employees want. The rewards system offers many opportunities and variety, here are some options:

- Rewards systems can vary, for example: For individual and group suggestion the reward is based on the score. Furthermore, for a group suggestion, the prize has to be shared amongst the group members.
- For every five suggestions (from thank-you to gold award) submitted within each fiscal year, the originator will receive an additional reward.
- Each year, the best five awards will get a special best suggestion of the year award – typically a plaque or other token of appreciation.

- Each department must submit their claim voucher to the administration department at the end of each month, and collect the cash awards.
- Other reward systems can have features such as:
- Gifts, tickets, dinner vouchers, etc.
- Accumulation of points during each year. At the end of each year, the total points accumulated can be used to collect awards.
- A percentage of the dollar savings in the first year, up to a maximum dollar value, of say $5,000. Some companies use such a reward system.

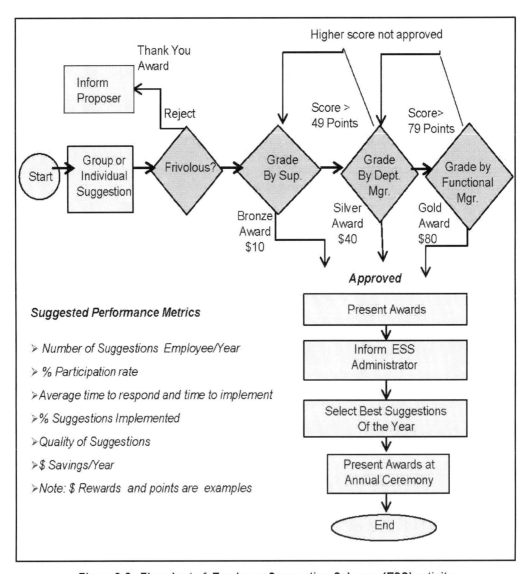

Figure 9-2 Flowchart of Employee Suggestion Scheme (ESS) activity

Guidelines for Supervisors and Managers: The immediate supervisor must decide whether the suggestion raised by the proposer is within the job responsibility or assignment. Suggestions that are within the direct control of the proposer will not qualify.

- Quick response time is important. A suggestion assessed within the proposer's own department should take less than a week. For a suggestion requiring another department to assess: two weeks.
- If assessment time exceeds guideline, the responsible supervisor must explain the reasons for the delay.
- The immediate supervisor can approve suggestions up to a defined $ value. The department manager can approve up to another $ level. All $ values above (say) $500 need approval by the functional manager.
- When computing "net savings" of a suggestion, only the first year savings will be considered.
- The immediate supervisor can assist the proposer when she makes a suggestion to resolve concerns which are beyond her capability.
- The immediate supervisor should reject all frivolous suggestions or suggestions outside the company's control, for example, "build a bridge across a river in order to reduce time taken to reach work".

Promotion activity: Constant promotion and encouragement will be necessary to make the suggestion scheme successful. In a separate box, we have given a random sample of the type of suggestions you can expect if you start a suggestion scheme. The promotion effort should focus on:

- Productivity gains not cash gains of employee.
- Giving awards in public at department, company meetings, or company functions.

Success factors for a suggestion scheme: From our discussion above and experience in running a suggestion scheme, we can mention a few:

- *Keep it simple*: Keep the suggestion-form simple. For operators allow a paper format, as operators may not have access to PCs in the factory. On the other hand, engineers and professionals will prefer a form accessible and usable on a PC; use the same simple form.
- *Give quick feedback:* Acknowledge the suggestion verbally or via email. Acknowledge within two days. Acceptance or rejection of the suggestion should be within a week in the same department and a maximum of two weeks across a company.
- *Monitor progress* and set metrics for success: here are some ideas based on the maturity of the scheme.
    - The Initial Phase:
        - Number of suggestions received and increasing trend.

- Percentage of employees participating.
- Time to respond to contributor.
- Time to implement suggestion.
  - The Mature Phase:
    - Yearly goals for suggestions per employee.
    - Quality of suggestions: Percentage of suggestions implemented.
    - Cash savings per year.
- *Recognition and engagement:* The most important factor is to focus on recognition and engagement with the employees. This will encourage participation and success.

What Type of Suggestions Can You Expect?

Suggestions will range from frivolous to excellent. With time and employee education, the quality and number of suggestions will improve. Here is a list of randomly selected suggestions that were taken from a multi-national company operating at several sites in Asia:

Sell, in order to recycle, waste material instead of discarding it. Result: Savings of $30,000 per year.

Eliminate wrong insertion of connectors, which causes 24 percent rejected connectors, by using a keyed connector. Result: Dollar savings were marginal, but rejects reduced to 0 %.

Modification of an assembly work-holder to increase production speed. Result: Savings of $12,000 per year.

Improvement of copper plating process. Result: Savings of $16,000 per year.

Propose a new application for a distance measuring instrument. Result: Increased sales of $550,000 per year.

Reduced number of copies of account payables invoices. Result: Savings of $5,000 per year and reduced paperwork burden.

This is indeed an impressive list. In addition to active employee engagement and participation, it represents improvements above and beyond what an organization's management would even think of doing. It helps enormously, then, if every employee of the organization keeps his or her eyes and ears open and suggests ways to improve products and processes resulting in increasing productivity, profits and customer satisfaction.

Education and Training

Training in the relevant standard work is necessary for all operators. But additional skills-training and education of all employees is crucial and necessary if a company is to succeed and achieve increasing productivity and customer satisfaction. Listed here are some basic methodologies that should be part of an effort to enhance employee skills.

Operators:

- Multi-skilled training in standard work. This includes OJT (On the Job Training) and certification.
- Training to use the seven quality control tools.
- Basic problem solving skills: The PDCA cycle and its applications. This will be followed by participation in kaizen projects, per department plan.
- Understanding of basic TPS/lean/TQM techniques. Example: Why balance the line, how to respond if your station is unbalanced and cycle time is not achievable; good training will encourage production staff to bring up issues when there is waste in their process.

Engineers:

- Hands-on training and education in TPS/Lean manufacturing skills. Understanding of all topics in this text.
- Training to use the seven quality control tools.
- Detailed problem solving skills. The PDCA cycle and applications.
- Statistics, design of experiments or Six-Sigma basics.

Managers and professional staff:

- Training and education in TPS/Lean manufacturing skills. Understanding of all topics in this text, including understanding practical applications.
- Detailed problem solving skills: The PDCA cycle and applications.
- If you have an annual planning process, similar to Hoshin Kanri, then training is required for new managers.
- Review of the competitive environment.
- TOC or constraint management theory: This is an excellent tool to communicate the understanding of constraints and how to resolve them in production and business systems.

Working with Partners and Suppliers

In addition to developing employees, we must also develop and work closely with our partners and suppliers. Only then can we expect a good, dependable, stable, and long term relationship.

Training and education of partners and suppliers: If you have a strong, World-Class suppler, such training may not be required. Even then they need understand company policies and direction. However, smaller but enthusiastic partners and suppliers should be invited to participate in internal training and education, similar to what we showed earlier for engineers:

- Hands-on training and education in TPS/Lean manufacturing skills, with emphasis on kanban and JIT manufacturing, quality expectations, kaizen, and other techniques they can learn from.
- Training to use the seven quality control tools.
- Detailed problem solving skills: The PDCA cycle.

Create a strong supplier base: Work with suppliers to improve, monitor their performance, conduct on-site inspection until they are certified, and provide feedback for improvement. Form joint company and supplier teams to resolve incoming quality and delivery issues. Work closely with them to ensure good cooperation and results; the teams should be given recognition for the results achieved. Once a supplier is certified, allow their parts to move automatically from 'dock to stock.' With a strong supplier base, the next step is to implement a kanban delivery process.

Implementing kanban with suppliers: In the chapter on kanban systems, we discussed the *constant cycle withdrawal mode*. This is the preferred method for running kanban with suppliers. This is due to the suppliers being outside the factory, often far away. The suppliers have to ship product over long distances; hence freight costs as well as traffic congestion become an issue. Therefore as demand changes, the order quantity changes but the delivery process remains constant. Most MRP systems can be programmed to trigger kanban requests to suppliers. Alternatively, you can apply the same process that we have discussed for your suppliers in the approach we used for implementing kanban in the factory. Certified local suppliers can deliver bulky and expensive parts daily to the production line; while distant suppliers can deliver parts more frequently. This will drastically reduce internal inventory and storage.

Summary: Employee & Partner Participation and Development

Employee development and participation and partnering with suppliers are very important elements for a successful company. A quest for operational excellence must have all employees contributing and engaging with the company. As we write, there is the case of one of the largest contract manufacturers in the World, which has reported numerous employee suicides[39].

It is in this light and perspective that we reiterate that an organization must develop, respect, and challenge all its employees. This will help create a stable and healthy workforce and an attractive work environment. We have discussed kaizen activity, employee suggestion schemes, and education as means to improve employee engagement and involvement. We reiterate that such *employee engagement and involvement is priceless.*

We also discussed the need to develop and work closely with suppliers. Only then can we expect a good, dependable, stable, and long term relationship. The result of all these activities will be higher employee morale, engagement, and productivity, and a strong and stable supplier relationship.

We have discussed many techniques and tools that will provide operational excellence and lead us to World-Class manufacturing. There is so much to do to get there; in the next chapter we discuss setting priorities.

Chapter 10
Doing the Right Things - Hoshin Kanri Planning

The manager does things right; the leader does the right things.
Dr. Warren Bennis

Overview

We have been discussing *how to do things right*: that is how to be productive, efficient, less wasteful, and have the best quality. However, it is equally important that we *do the right things*. This requires creativity, insight, and good planning. In today's highly competitive and changing marketplace the margin for error is decreasing, hence planning and executing the plan are key activities in a company. Planning also provides other benefits, including sharper objectives, improved performance standards, and management involvement. All of these result in a planned approach to tackling the market place that can have higher productivity, quality, sales, and profits.

In this chapter we will discuss:

- The essential elements of the *Hoshin Kanri* planning process, also called Policy Deployment. This is a complete planning process with objectives, strategies, implementation plans, and progress reviews.

- A short discussion on an alternative planning process using TOC (theory of constraints) methodology.

- Doing the right things also demands leadership; we will discuss the essence of leadership in the lean manufacturing environment.

Making the Future Happen via Planning, Execution, & Control

According to Charles Knight, Ex-CEO of Emerson Electric[40]: *"We believe that companies fail primarily for non-analytical reasons: Management knows what to do but, for some reason, doesn't do it."* The future can be shaped through careful planning and execution. Planning is one way to do what needs to be done. We will discuss how to keep plans focused, manageable, and successful.

Essentials of a Planning Process

A good planning process needs to exist throughout an organization. It should be a standard process with everyone trained in it. A good planning process will have a long range, strategic, plan and an annual plan. Our discussion, however, will look at the annual planning process. Such an annual plan will include an analysis of the current situation, the objectives that must be pursued, actual implementation, and progress reviews. This plan must focus on breakthrough product and service strategies that will lead to market success. We recommend the use of the *Hoshin Kanri*[41] planning process.

Hoshin Kanri Planning

This is a very successful planning process used by most Japanese Companies that have won the (Japanese) Deming prize for Quality. It originated in the 1960's at the Bridgestone Tire Company in Japan. At that time, weaknesses in their planning process were becoming apparent. The Hoshin Kanri philosophy originates from ancient Japanese military traditions and efficiency. Most Japanese companies and several American companies have adopted Hoshin Kanri and Nichijo Kanri type planning.

The terms *Hoshin Kanri* and *Nichijo Kanri* are loosely translated from the Japanese as follows:

Hoshin Kanri: Hoshin means objectives or directions, while Kanri means control or management. So in essence Hoshin Kanri means Policy Management or Management of Objectives. Henceforth let us just call it the Hoshin plan.

Nichijo Kanri: Nichijo means Daily. Hence Nichijo Kanri means *daily management*. In the Western corporate environment a clearer definition could be KPI (key process indicators) or Business Fundamentals. We will use the terms interchangeably, but prefer the term Daily Management Plan.

A description of the Hoshin planning process is given next. In the succeeding pages the various formats to be used are described and suggestions to develop a good plan will be given with examples.

The Hoshin Kanri Planning Process

The Hoshin Kanri process follows the PDCA cycle. The process begins with a review of the following items (refer to the planning flow in Fig. 10-1):

Company or organization vision: The Company's or Organization's vision is reviewed, discussed, updated, or improved. The annual plan must be in harmony with the vision.

Long-range plan: If the long-range plans are established at the company or organization, these are reviewed and the relevant strategies are incorporated into the annual Hoshin plan.

Specific customer inputs: All recent and significant customer inputs, issues, and needs that have been collected over the past year are reviewed. In addition, ongoing pressing issues must also be reviewed. Such data can come from routine customer meetings, visits, letters, surveys, and from customer complaints. The need for closeness to customers can never be overemphasized.

Current economic situation: The current economic situation is reviewed and the necessary response discussed and planned for.

Review of the previous year's plan: Successes and failures in the previous year's plan are reviewed at all levels in the organization. The lessons learned will influence the upcoming year's plan.

All of the above items form a list of key issues from which the Hoshin plan will be prepared. Typically, the entity General Manager (which can be the CEO, President, or a product or service division manager) will prepare the annual Hoshin plan. This plan will provide broad objectives, strategies and performance measures. The next level managers – for example, marketing, finance, research and development, operations or manufacturing, and quality – will then prepare a more specific plan, after discussions with their department managers. In addition the department managers will prepare an implementation plan. This process of management discussions to finalize the plan is often called "catch-ball" – which comes from the term used in baseball. Refer again to the Hoshin flow chart in Fig. 10-1.

Daily Management Plan

So far the Hoshin plan covers the objectives that flow down from upper level management. These indicate what the entity must achieve in the year. But what about day to day items or items that require both controlling and monitoring? For example: Cost controls, employee morale, process of selling or manufacturing a product, controlling a product or process, or some other annual repetitive task.

The Daily Management plan (sometimes called Business Fundamentals plan or KPI – key process indicators) addresses this issue and focuses on controlling or "keeping the house in order". We prefer the term Daily Management plan because this plan is used to monitor and manage the day-to-day metrics that are important in an operation; it tracks annual repetitive tasks. There is very little higher management level involvement in preparing this plan, but all levels of management should have it. The final plan should have both Hoshin and Daily Management plans.

Launching the Plan

Next, the plan is launched. Refer again to Fig. 10-1. Often a facility wide meeting is held to present the plans, but at a minimum the plans should be presented to all management and supervisory staff.

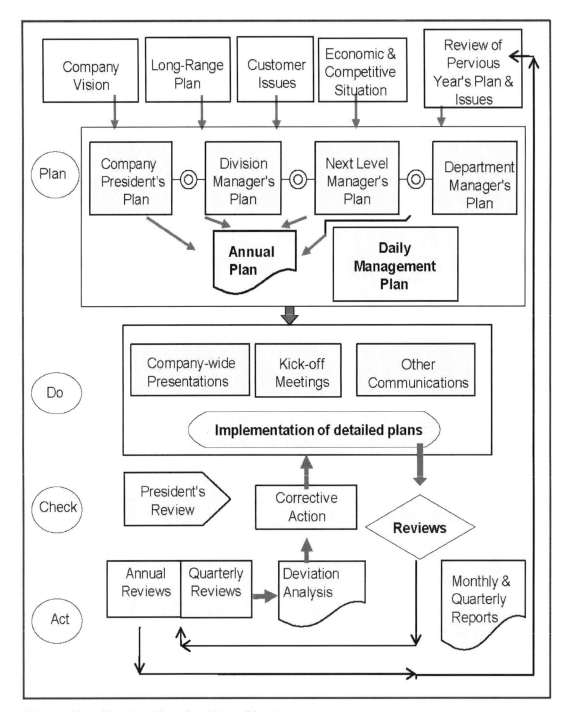

Figure 10-1: Hoshin Planning Flow Chart.
This chart shows the sequence of activity for generating a Hoshin and Daily Management plans. Note the catch-ball symbol (◯) which denotes management discussions.

Doing the Right Things - Hoshin Kanri Planning   197

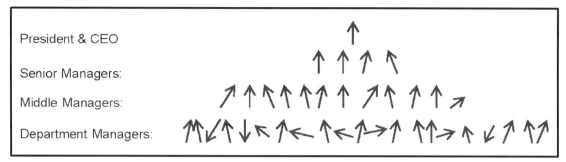

Alignment of Objectives in Typical Planning

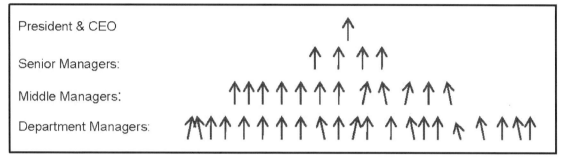

Alignment of Objectives in Hoshin Kanri Planning

Figure 10-2 Illustrating the difference between Typical and Hoshin planning

Hoshin Plan Reviews

Hoshin plan reviews are held regularly – monthly or quarterly. Many operations prefer to do them monthly, which we recommend. If everything is on track, it's business as usual. If not, plans have to be modified and corrective actions taken to address the root cause of problems, and these become additional tactics in the plan. Finally, the results and experiences of the current year are summarized in an annual Hoshin review. The annual review is done during the last month of the planning year; this sets the stage for starting the next year's annual planning cycle. Reviews are done at all levels, wherever there is a Hoshin Plan owner: Be it CEO or manager.

Illustrating Hoshin Planning

In Fig. 10-2, we illustrate the benefits of Hoshin planning. The figure shows how the use of Hoshin Kanri planning ensures better linkage and alignment of objectives at the various levels of management in the company. This is due to the Hoshin planning methodology, which ensures very tight cascading of objectives from one management level to the next. In fact the original Kanji (Japanese) characters of Hoshin Kanri mean "shiny needle or compass"; hence Hoshin Kanri planning directs everyone in the organization towards the same direction

Relationship Between Annual Plan and Daily Management/Business Fundamentals (KPI)

Annual Plan:
Directing Your Company (Battleship) to its new market or destination, while ensuring a safe journey by avoiding attacks, enemy aircraft, torpedoes.

Daily Management Plan/Business Fundamentals or KPI:
✓ Keeping the battleship well maintained and running smoothly
- Maintaining the ship
- Checking fuel, engines, regular cleanup

✓ Feeding, keeping healthy, and entertaining the ship's crew

Figure 10-3: The relationship between an Annual Plan and the Daily Management Plan

ANNUAL HOSHIN PLAN			
Objective	Target/Goal	Strategy & Owner	Performance Measure

Figure 10-4: Format of Annual Hoshin Plan

Hoshin Plan: Ensuring Success

The strength of the Hoshin and Daily Management planning is due to a systematic and tightly coupled process. This process ensures three things: First that the management team has committed to it. Second, a hierarchy of objectives and strategies will occur across the organization. Third, and most important, the plan is reviewed regularly and corrections made to get the entity or ship back on course.

Relation between Hoshin Plan and Daily Management Plan

At this juncture, it will be useful to discuss the difference and relationship between Hoshin Kanri Planning and Daily Management Planning. We illustrate the difference in Fig. 10-3.

Imagine a Battleship (or Company) sailing to its new destination (or Market). The annual Hoshin plan ensures it gets to its destination safely, while avoiding (competitive) attacks and other undercurrents. But its Daily Management plan looks at controlling or maintaining the ship, checking and replenishing fuel, running the engines, providing regular cleanups; it also looks at feeding, sustaining, plus entertaining the ship's crew. Obviously you need both types of plans and the separation into Hoshin Kanri and Daily Management is very useful for directing management focus, allocating appropriate resources, and setting priorities.

Formats and Guidelines

Next, let us review some recommended guidelines and formats for the Hoshin planning process and Daily Management. More details on planning formats and templates are provided in the references within the appendices.

Annual Hoshin Plan

The Annual Hoshin Plan summarizes the breakthrough objectives for the entity or organization. These could be objectives that let you leapfrog your competition. These objectives normally require more than the ordinary sustaining effort to accomplish and are likely to involve multi-department collaboration. The plan generally comprises 4 elements:

1. Objective
2. Target or Goal
3. Strategy
4. Performance Measure

A recommended format to capture these elements is shown in Fig. 10-4, followed by an explanation of each element. Completed examples with comments, are shown in Fig. 10-5 and 10-6. An explanation of the elements is appropriate:

- *Objective:* This is the aim to be achieved: An aggressive or breakthrough statement.
- *Target/Goal:* This is a broad indicator measuring accomplishment of an objective. It must be established for every objective and be quantifiable.
- *Strategy:* This describes the method of achieving the goal.

- *Performance measure:* This is used to determine the progress or completion of a strategy. It consists of a statement and number indicating the target to be achieved.

Let us now examine some of these elements in detail.

Objective

Before objectives are prepared you should develop an issue list. Refer to the Hoshin planning flow chart in Fig. 10-1. This starts with a review of the following:

- Company vision and the long-range objectives.
- Specific customer inputs and issues.
- The current economic and market situation.
- New areas or direction for the company.

Successes and failures in the previous year's plan. This point is extremely important – successes and failures of the previous year are reviewed and analyzed before a new plan is prepared. In an adjacent page and box, entitled "An Example of an Issue List from the Previous Year", we show an example.

These objectives will determine the direction in which your business is heading. As a guide, the plan cover four categories, abbreviated as QCDE:

- *Quality (Q):* This includes customer satisfaction and product/process quality
- *Cost (C):* This includes all costs, such as, administrative expenses, manufacturing costs and productivity issues.
- *Delivery (D:* This includes new product design introductions, R&D and manufacturing product commitments, and delivery of products to the customer.
- *Education (E):* This includes human resource training and education.

The concept of QCDE is meant to ensure that nothing important is overlooked. We do not suggest that you have four objectives, we only suggest you review all these four categories. In fact we recommend that senior managers do not have more than two objectives – with several supporting strategies. In actual practice, some of the objectives can have strategies that address more than one of these categories. For example, an objective to increase profits can address costs, delivery, and quality. Refer to the example in Fig. 10-6. There are several points to be made on the concept of QCDE:

1. The QCDE categories are considered essential for business success; progress in each category will help keep a company robust, healthy, and competitive. Complacency in any category may cause problems. After generation of the key issues, they can be categorized in the QCDE categories.

2. Ideally, the number of objectives should be limited to no more than two. This may be difficult but can be achieved if a manager focuses only on "breakthrough" objectives – that is, objectives that are essential for success. Experience has shown that an individual manager can only focus on one or two objectives in a large organization.

3. In any year, not all the QCDE categories need to be addressed. For example, if in the previous year you set up a comprehensive skills-training program, you would not worry about it in the current year because it is now a Daily Management item. The Human Resources Department is managing this on a day to day basis. However, the delivery or new product category would certainly be a yearly issue.

4. Often, the QCDE categories will merge. For example, a profit objective could include strategies for Quality, Costs and Delivery – refer to Fig. 10-6.

5. In setting objectives, a hierarchy is important. So in a service organization, the general manager could look at the Q category and set a top level customer satisfaction objective with a goal being the result of an industry wide survey. But at the lower level a product or service manager would have a specific objective that supports the top level objective and goal – say an objective to improve a product or service that customers are unhappy about. A similar analogy will apply for the other QCDE categories.

6. The C or Cost Category is often overlooked in many organizations. This often becomes a concern only just before or after a company takeover, acquisition, management change, or a sudden downturn in profits. When we mention costs, we do not mean just specific product or service costs. *We mean all costs.* This includes overhead, manufacturing, and quality costs which have a tendency to creep up, especially during good times. Ignoring costs – that is, not managing them on a year to year and a daily basis – can result in a *self-feeding cycle of competitive decay.*

FY 2011 ANNUAL HOSHIN PLAN:

Prepared By: GM		Division/Department: Apex Div		Page 1 of 2

Situation Analysis: Apex Div. profit has been declining. This has been due to an old product line, high cost structure, and delayed introduction of products. New products (Objective 1.) will increase sales and profits. However, we must reduce current costs by 35 %; in addition field failures must be reduced. In all this will boost profits from 3.5 to 6%.

OBJECTIVE	No	STRATEGY	Owner	PERFORMANCE MEASURE
1.0 Dramatically Increase Revenue	1.1	Introduce products for new customer segments (R&D & Sales)	R&D Mgr.	*Complete design and ship WFP products by May. *Provide 40% of sales in 2011
	1.2	Improve the supply chain in the areas of Reseller training and inventory reduction. (Marketing)	Mkt. Mgr	*Resellers rate us as the number one company to do business with.
TARGET/GOAL Orders at $ 1.5 Billion	1.3	Improve/Refresh current products offerings to get a sales increase of 20% (R&D)	R&D Mgr.	*Improve product X per customer request and release in May. *Sales increase of 35 %
	1.4	Develop a system strategy to manage current growth in electronic commerce, links to resellers, and new needs (IT)	IT Mgr	*Rollout solution for E-Commerce growth by April, 2010

Comments on these plans:

* The Gen. Manager develops his plans as shown above. The appropriate next level manager picks up whichever strategy is assigned to him/ her and makes it his/her objective.
* In this example the sales & marketing manager picks up the general manager's strategy no. 1.2, which becomes his/her strategy and its performance measure becomes the goal.
* In a similar fashion, all the Gen. Manager's strategies will be deployed to the next level.

FY 2011 ANNUAL HOSHIN PLAN:

Prepared By: Marketing Manager		Division/Department: APEX: Marketing		Page 1 of 2

OBJECTIVE	No	STRATEGY	Owner	PERFORMANCE MEASURE
1.2 Improve the supply chain in the area of Reseller training, inventory reduction, and E-Commerce linkage.	1.2.1	Provide training to current channel partners in the areas of increasing product demand, promotion management, after sales support, and inventory management.	Chanl Mgr	*Training for Resellers by 1Q * Increase Channel sales by 35%
	1.2.2	Work with the IT operation to develop a system to automatically track both channel inventory and obtain real-time product sell through information.	Chanl Mgr	*System pilot by 1Q, and implementation by April
TARGET/GOAL Resellers rate us as the number one company to do business with.	1.2.3	Set up a process to track and manage reseller inventory and keep it at optimum levels.	Mkt. Mgr	Inventory reduction in the channel of 40% by 3Q.

Figure 10-5: Preparation and deployment of a Product Division Manager's plan.

An Example of an Issue List from the Previous Year

In preparing a Hoshin plan, you need to do a review of the previous year's performance, at the end of that year. What can you expect from an end of year review? It would be a list of problems, or lessons learned from a job well done. This list of issues can influence the following year's Hoshin plan. Here is a list from product divisions doing sales, design, and manufacturing. Note: they were compiled from different product divisions.

- Our profitability was below target; we need to address the root causes and improve the situation.
- Planned new product family X-22 is behind schedule and is impacting revenue and profitability. The R&D program and team is not performing. Why?
- The new product seems to have been introduced in a hurry: Failure rates in the factory, field, and customer complaints are high. Did we use DFM and Process-FMEA techniques?
- Lost 30% of big deals when competing for a sale of product X12. There is a need to understand what happened: Is it a product or marketing problem?
- Sales representatives and engineers are inexperienced due to extensive hiring during high growth. Need training as soon as possible.
- High inventory of the Z33 product in manufacturing. Market forecasts may be the cause or the build to order process is broken, we need to investigate and resolve.
- We received excessive customer complains for Product C234, field defects are very high and the issue has not been resolved. Marketing says the product is still selling well and sales can increase if we fix the problems quickly.
- Overall manufacturing yields on Product A22 (internal and at outsourced supplier) have not met target and are reducing profitability. Yields on Product B28 are exceeding targets. What are lessons learned?
- Won 95 % of big deals when competing for a sale of product Y22. Why did we do well? Can we transfer the lessons learned to other new products?
- Our latest product is doing well and has exceeded all expectations. What did we do well? How can we translate this into lessons learned for new business?

FY 2011 ANNUAL HOSHIN PLAN:

Prepared By: GM		Division/Department: Apex Div.		Page 2 of 2
Situation Analysis: Apex Div. profit has been declining. This has been due to an old product line, high cost structure, and delayed introduction of products. New products (Refer to Objective 1.) will increase sales and profits. However, we must reduce current costs by 35 %; in addition field failures must be reduced. In all this will boost profits from 3.5 to 6%.				
OBJECTIVE	**No**	**STRATEGY**	**Owner**	**PERFORMANCE MEASURE**
2.0 Increase profits	2.1	Revamp Service business, increase throughput and lower costs. (Marketing)	Mkt. Mgr	*Increase service business profits from 4 to 8%.
	2.2	Reduce costs in Operations, Admin, Mktg. (Ops, Admin, Mktg.)	Fun Team	*Cost reduction of 30%
TARGET/GOAL	2.3	Reduce External Product field failure rates to reduce warranty costs. (Operations)	Ops Mgr	*Reduce failures from 4 to 2%
Profit increase from 3.5 to 6 %	2.4	Accelerate JIT/Kanban processes and reduce WIP and overall Inventory. (HR:training; Operations: rollout)	Ops Mgr	*Inventory reduction by 25% plus eliminate all external warehouses by June

Comments on plans:
The GM's has selected profit as her no. 2 objective. This is a corollary of C in the QCDE categories. The strategies for the objective include Q, C, & D items.

* The Operation Manager selects the strategies that are deployed to him.

* Note the numbering system provides traceability of objectives & strategies: The GM's objective is no. 2.0, and his strategies are 2.1, 2.2, etc.; The Operation Manager's objectives take on the number of the GM's strategies: 2.2 and 2.3; and the strategies become 2.2.1, 2.3.1, etc.

FY 2011 ANNUAL HOSHIN PLAN:

Prepared by Ops Manager		Division/Department: Apex Ops		Page 1 of 2
OBJECTIVE	**No**	**STRATEGY**	**Owner**	**PERFORMANCE MEASURE**
2.2 Reduce manufacturing costs 2.3 Reduce external product failure rate	2.21	Review and reduce procurement costs for all products with life of more than 8 months. (Procurement/Materials manager)	Proc. Mgr	*Material Cost reduction for product A, B, C of 20% by end 2Q.
TARGET/GOAL 2.2 Cost reduction: 30% 2.3 Fail rate reduction: 4 to 2%	2.22	Redesign and simplify product A and B. These products have life of 2-3 more years.	R&D Mgr	*Reduction in assembly time of product A, B of 15% & 23% by end 3Q
	2.23	Increase outsourcing of non-critical sub-assemblies by April.	Proc. Mgr	*Target for 50% increase in outsourcing. Target for 30% cost reduction.
	2.31	Reduce failure rate of products B, C, & D.	Ops Mgr.	*Reduce failures by 50% for shipments beyond June.

Figure 10-6 Preparation and deployment of a General Manager's plan.

Chief Executive's Objectives

We have given some guidelines for selecting and grouping objectives, basically using the concept of QCDE. If you are managing a complex organization – say several divisions or several countries – then it may get more difficult to find a common denominator. In addition, you must be strategic and not have too many objectives.

What is too many? More than two is too many. Experience has shown that one or two breakthrough objectives are about the maximum that an organization can manage and achieve.

Another good reason for the chief executive to have few objectives is that it will allow lower-level managers to add objectives that are specific to their situation. Otherwise you have a situation where the chief executive has, say, four objectives and the general manager adds a few more. Very soon the operation is awash in plans and objectives, resulting in a loss of focus.

Our recommendation is for the chief executive to follow these guidelines in setting objectives:

1. *Look for common issues or challenges.* For example, in the area of new products, growth, and profit.
2. *Review specific needs of a weak area in the organization.* For example, an underperforming product line, division, or country.
3. *Review future needs of an organization.* For example, startups in new countries, new business, or new technologies.

Strategies and Performance Measures

Let's review strategies and performance measures.

Strategy: A strategy describes the procedure by which the targeted goal is to be accomplished. An average three to five strategies is recommended for an objective. However, the actual number will vary depending on the complexity of the objective.

Is there a technique for generating strategies? Yes. We recommend two methods that we have used successfully. But before we offer our ideas, here are observations on how several brilliant CEOs have done product strategies in the past: At Hewlett-Packard, Bill Hewlett requested a scientific electronic calculator that could fit in his shirt pocket, resulting in the top-selling HP35, which effectively eliminated the slide-rule. At Sony, Akio Morita requested the popular Walkman music player, which started a generation of personal music players. More recently at Apple, Steve Jobs came up with the stunning I-Phone. Our techniques, however, are more ordinary and are given here:

Generating strategies, method 1:

1. List the objective with its goal.

2. Generate a list of strategies (by brainstorming) which can help achieve the goal. Typically this will involve understanding of constraints, roadblocks, technology needs, and success factors.
3. Evaluate the choices by ranking each according to its contribution towards fulfilling the objective, its cost, its feasibility, and other limiting factors.
4. Establish a ranked order of choices.
5. Select the choices based on ranking; the assumption is that we may not be able to do everything due to resources.
6. Discuss the choices with relevant staff. The catch-ball symbol shown in Fig. 10-1 indicates discussion nodes between staff. The term catch-ball originates from baseball terminology, where the outfielder may throw a ball to the pitcher. Both players must ensure that the ball is caught successfully.

Generating strategies, method 2:

A second method of generating strategies is a more scientific and data driven method. As we have mentioned previously, the Hoshin process is a large PDCA cycle – and preparing the initial plan is the P stage of this cycle. This is what we need to do:

1. After the objective and goal is determined, prepare a cause and effect diagram – where the effect is the selected objective.
2. Many of the causes can be determined via brainstorming.
3. Determine the most likely causes that affect the objective and verify with data.
4. Convert the verified causes into strategies.
5. Check and be confident that, after implementation, these strategies will add up to meeting the objective and goal.

Let us review an example of how this is done. If the objective is: "to increase profitability". Then a cause and effect diagram for "why are profits low?" can be constructed. An example is shown in Fig. 10-7. Next the most likely causes are selected, discussed, and verified. These have been verified and marked in the cause and effect diagram, and are:

- Lack of new products
- Product failure rate too high, although at target
- High manufacturing and administrative costs

Appropriate strategies can now be prepared to achieve the profitability objective and goal. This is shown in the Hoshin plan in Fig. 10.6. In this example we expect that the strategies to add up to meeting the objective and goal. Look again at Fig. 10-6, here the situation statement at the top of the form indicates a financial analysis was done and specific performance measures were selected to ensure the goal would be met if the performance measures were met.

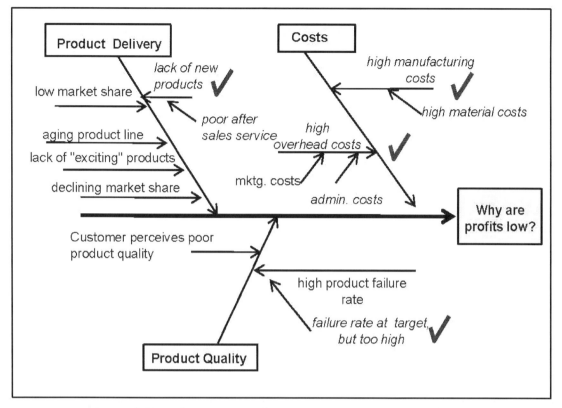

Figure 10-7: Cause and effect diagram for low profits

Performance measures and types of performance measures: A performance measure is used to determine the progress or completion of a strategy. It consists of a statement and a number; the number indicates target to be achieved. There are two types of performance measures:

- *Result-oriented or end-of-process performance measure:* This should be a way to measure the outcome or desired result of the strategy; for example an action plan, increased sales or higher production line yield.
- *Process-oriented on process tracking performance measure:* This should be a way to measure progress of the strategy; for example, phased results, interim action steps, or targets for each step of a process.

Situation analysis: In preparing objectives and strategies a detailed review of the current situation is important. In the Hoshin flow chart in Fig. 10.1, we show the various items that we must review before we complete the plan. Such a detailed analysis is crucial in helping us to select the few breakthrough objectives that are required. In the analysis we should be able to determine the appropriate goals and performance measures we wish to select. In Fig. 10-6, we show a Hoshin Plan with a brief situation analysis.

Deployment and Cascading of an Objective

Deploying an objective in a large organization: Deploying an objective in a large organization is easy and effective with the Hoshin planning process. Certainly it is much easier than with any other planning process that we know of. This is how the objective gets deployed: The General Manager's plan will be completed and deployed to the next level. The next level managers will select the strategies that are appropriate to them – these become their objectives and the performance measures becomes their targets or goals; each of these objectives will generate a number of new strategies. The concept is illustrated in Fig. 10-8.

But not all strategies cascade down the organization. Some strategies may not be deployed and are picked up at a high level – we show this in Fig. 10-12. Here the Marketing Manager deploys her own strategy (from Fig. 10-5) into an implementation plan. Other strategies may involve many departments and are best assigned to a cross-functional team led by a specific manager. For example a strategy for a new product will involve R&D, marketing, quality, and manufacturing. More on cross functional strategies later in this section.

For the process to be effective, the Chief Executive's plan has to have one or two objectives – because at each lower level the objectives will multiply. If done well, the result will be a tightly knit plan across the company, with everyone moving in the same direction.

Impact of cascading and how to prevent objectives from repeating: As Hoshin plans cascade down the organization, a number of things – good and bad – will happen. The number of objectives will quickly multiply, but if they are prepared well, there will be strong linkage and synergy, resulting in a sharply focused organization. That is why we recommend starting with one or two objectives at the top. Alternatively, if top-level objectives are too tactical, objectives will start to repeat. Here are some guidelines:

1. *Problem:* Objectives repeating because they are too tactical. For example, if we have a five-star general or chief executive with a very tactical plan such as "Take the hill" or "Hire five sales executives", then the cascading process will not work; the result will be repeating objectives at many levels.

    Solution: Because of the nature of the cascading process, the top level objective and supporting strategies must be broad and as high level as possible. If this is done the lower-level objectives (and supporting strategies) can be more tactical.

    For example:

    At the highest level we have "Win the war" or "Increase profits"

    At the next level we have "Destroy Atlantis", or "Add a new sales force", or 'Improve product quality".

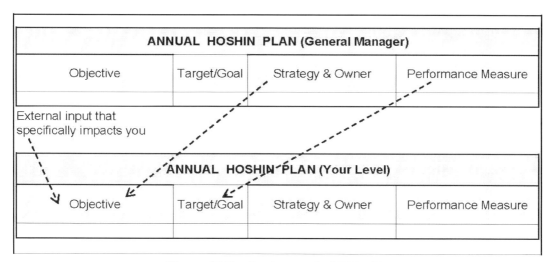

Figure 10-8: Deployment of Objectives

2. *Problem:* Objectives (and supporting strategies) repeat at different levels if there are too many layers of management in the organization.

   Solution: This is more difficult to resolve but may indicate there are too many management layers in the organization. If the layers cannot be reduced, then request the lower layers to use implementation plans.

3. *Best method:* The best way to get good cascading of plans, which are compact and concise, is to stipulate only a few layers of Hoshin plans in an organization. These are then supported by implementation plans.

Hoshin Plan Deployment Matrices

Matrices to deploy plans, identify responsibilities, and communicate effectively are often used in Hoshin planning. We show two types here. A frequently used matrix is the Hoshin plan deployment matrix. This is useful to ensure effective deployment of higher level objectives. This is shown in Fig. 10-9. Included in the matrix is a list of cross-functional and kaizen teams to mange specific strategies. These teams will present their progress during Hoshin plan reviews.

It is crucial that the management team determine which strategies require cross-functional teams consisting of representatives from different functions such as R&D, marketing, manufacturing, and so on. In the strategy in Fig. 10-6, the strategy of revamping the service business requires an entity-wide team effort. The cross-functional team will:

- Be led by a manager, with representatives from all necessary functions.
- Assign responsibilities, tactics, and specific goals for staff or departments.

- Meet regularly to measure progress, review resources, and set priorities to ensure the strategy is successfully completed.

Hoshin Deployment Matrix					Division: Apex			Page 2 of 2	
Functions	R&D	Sales	Marketing	Finance	Operations			Related Activity	
Objective 2.0: Increase profits Strategies:					Materials	Eng.	Prod.	Cross-Functional Team	Project Team
2.1 Revamp Service business, increase throughput and lower costs.	0	X	X	M	M	X	0	Leader: Mktg. Manager	None
2.2 Reduce costs	0	0	X	X	X	X	X	None	5 Teams,
2.3 Reduce external product failure	0	0	0	0	X	X	M	None	1 Ops Team
2.4 Accelerate JIT, Kanban processes and reduce WIP, and overall Inventory.	0	0	0	M	X	X	X	None	1 Ops Team
KEY:	X = high relationship, 0 = no relationship, M=medium relationship								

Figure 10.9 A Hoshin Deployment Matrix

This matrix can be used to deploy an entity manager's objective and strategies. The matrix can be used to determine who picks up the various strategies that were shown in the GM's plan, Fig. 9-6. This way there will be little confusion and strategies and performance measures can be better formulated. Note the matrix also lists strategies that require cross-functional teams, and those can be managed by department-level teams.

Alternative Hoshin Planning Format and Deployment Matrix

We have shown the Hoshin plan format (Fig. 10-4) and the Hoshin plan deployment matrix. We concede from our experience that this will result in many forms and paperwork. One effective solution is to use a web-based electronic system to manage the paperwork, reminders, reviews, and so on; check the references for information on the PlanBase System. We also provide an alternative format for Hoshin planning and deployment. This is shown in Fig. 10-10. This matrix cuts paperwork and eliminates many layers of plans. The matrix is very comprehensive and shows:

- Summary list of long term objectives (2-3 year time frame).
- Summary list of Annual objectives.
- Corresponding strategies of Annual objectives.

- Performance measures/targets for the above strategies.
- Strategy deployment: Strategy owners, department strategies, or cross-functional strategies.

In the matrix, we have included the General Manager's two objectives and strategies from Fig. 10-5 and 10-6. This matrix can be used for the entire Hoshin planning process. Just one matrix to list everything: Long term objectives, annual objectives, strategies, performance measures, and strategy deployment and owners. Let's look at how one objective is managed with this matrix:

- Objective 2.0 is to increase profits: One of the strategies (2.4) is to accelerate JIT/Kanban and reduce WIP; the performance measure is a 25% reduction in inventory. This is assigned to Operations (Production and Materials). The Operation Manager will have to assign a manager to lead the strategy and prepare an implementation plan.
- A second strategy (2.2) is to reduce costs, with a performance measure of 30% cost reduction. This is deployed to Finance, Operations, and Marketing. A cross-functional team will be required and led by one manager. The team will need an implementation plan.
- Hence we have ONE Hoshin plan and many implementation plans.

Hence the question: Why not use this format instead of the one we proposed in Fig. 10-4? Yes, you should first consider this alternative format. However, this is more suitable for a smaller organization which is not geographically dispersed or where the management team is small or concentrated.

Still, there are some disadvantages to this format:

- Due to the need for short statements important clarification is left out.
- There is less opportunity for cascading of detailed strategies, as was done in Fig 10-5 and 10-6.
- Here only the General Manager's objective and strategies are shown. Often a comprehensive plan needs more strategies to allow for a detailed plan that covers all needs. For example the strategy to improve the supply chain in Fig 10-5 requires three detailed strategies (at the next level) to achieve the goal.

Nevertheless, there is room for this alternative Hoshin planning matrix in smaller organizations and it can be used initially, unless the more detailed planning format is preferred. We have used both Hoshin long-form and short-form formats. The short-form has a lot going for it.

Regardless of which format you use, you will still require implementation plans and Hoshin review plans: Often these are provided in tabs in an Excel file in the alternative Hoshin plan matrix. But they need to be as detailed as the procedures and formats we show next.

Hoshin Plan & Deployment Matrix 2011			Division: Apex						Prepared by GM				Date: Jan 1, 2011					
X		2.4) Accelerate JIT/Kanban processes, reduce WIP, and Total Inventory							X						X	X		
X		2.3) Reduce External Product field failure					X					X		X				X
X		2.2) Reduce costs in Operations, Admin, Mktg.						X				X		X	X	X		X
X		2.1) Revamp Service business, increase throughput and lower costs.				X						X		X				
	X	1.4) Develop system strategy to manage growth in e-commerce, links to resellers, & new needs				X						X						X
	X	1.3) Improve current products offerings			X					X	X	X		X	X			
	X	1.2) Improve the supply chain: Reseller training and inventory reduction		X								X			X			
	X	1.1) Introduce products for new customer segments	X	X						X	X	X		X	X			
		Improvement Strategies Projects																

Annual Objectives & Goals		Long Term Objectives
2.0 Increase Profits: Goal 6%	1.0 Increase Sales: Goal $1.5B	Sales at $4 Billion
		New Products Per 3-Year Project Plan
		Market Share > 45%
		Number 1 in Customer Satisfaction

Specific Goals / Performance Measures:
- Complete design and ship WFP products by May.
- Provide 40% of sales in 2008
- Resellers rate us as the number 1 company
- Improve product X per customer request and release in May.
- Rollout solution for E-Commerce growth by April, 08
- Increase service business profits from 4 to 8%
- Cost reduction of 30%
- Reduce failures from 4 to 2%
- Inventory reduction by 25% + eliminate external warehouses by 6/08

Owners & Deployment Level:
Sales | Marketing & Service | R&D | Operations | Engineering | Production | Purchasing & Materials | Human Resources | Finance | Information Technology

Figure 10-10: Alternate Hoshin plan and deployment matrix

Implementation or Project Plan

Hoshin plans may cascade down for several layers of management. At the last layer there needs to be an implementation plan: This plan lists the detailed steps necessary to accomplish a strategy in the Hoshin plan. Implementation plans are typically prepared at the lower management levels or by professionals – basically these are the staff that must ensure the plans are successfully completed. Often a senior manager may have an implementation plan if he is the owner of the strategy. The implementation plan format is shown in Fig. 10-11; the following is a discussion on each element.

- *Strategy with performance measure:* Each strategy from the Hoshin plan is listed here, with the corresponding performance measure.
- *Implementation details and timeline:* The rollout of each strategy should be planned in detail. There are some choices:
  - *Gantt chart format:* List all the steps that need to be done for the strategy to be completed plus provide a timeline.
  - *PDCA format:* List all the steps in the PDCA cycle

The benefits of a detailed implementation plan are:

- Every Hoshin strategy in the organization ends up as a detailed implementation plan, with owner and timeline.
- The implementation plan is the evidence that the Hoshin plan strategy has cascaded down into actual work. When the implementation plans become part of the company staff's daily work plans and calendars, then the chances of success become very high.
- Progress can be measured by checking against the timelines, and deviations can be spotted and corrected. This provides an excellent format to review a manager's or employee's progress.

Fig. 10-12 shows how the Marketing's Managers Hoshin plan, from Fig 10-5 cascades into an implementation plan. This format is different from Fig. 10-11, as we have used a web-based Hoshin Planning format, courtesy of PlanBase Inc.

			IMPLEMENTATION PLAN					
No.	Strategy and Performance Measure	Implementation Details	Timeline					Owner
			Jan.	Feb.	Mar.	Apr.	-----	

Figure 10-11: Format of Implementation Plan

1.0: Dramatically Increase Revenue
1.2: Improve the supply chain in the areas of Reseller training and inventory reduction. (Marketing)
FY 2011 1Q Review 1.2.3: Set up a process to track and manage reseller inventory and keep it at optimum levels.

Business Unit	Function	Plan Owner	Plan Year(s)
APEX	Marketing	General Manager	Jan 2011 - Dec 2011

Situation Analysis

Apex Div. has been declining. This is due to an old product line, high cost structure, and delayed introduction of products. New products will increase sales and profits. However, we must reduce current costs by 35% in addition field failures must be reduced. In all this will boost profits from 3.5% to 6%.

Objective/Strategy (Owner)	Metrics (Actual / Target)	Status	Analysis of Deviation	Corrective Action
1.2.3: Set up a process to track and manage reseller inventory and keep it at optimum levels. (Marketing Manager)	Inventory reduction in the channel of 40% by 3Q: (15 / 15)	○↑L	On track with new improved process coming on line April 4th.	

Tactic (Owner)	Jan 11	Feb 11	Mar 11	Apr 11	May 11	Jun 11	Jul 11	Aug 11	Sep 11	Oct 11	Nov 11	Dec 11	Status St-Cu-Fi	Remarks
1.2.3.1: Value Stream Map the inventory management process (Marketing Manager)	3-21												✓	Value stream mapping yielded excellent improvement opportunities
1.2.3.2: Streamline inventory management process (Marketing Manager)	7	-18											○○	Streamlining completed and offers big improvement
1.2.3.3: Implement new tracking inventory process (Marketing Manager)		28											⊗⊗✓	Fixing one small bug extended schedule by two days
1.2.3.4: Implement new limits on reseller inventory (Marketing Manager)				4										
1.2.3.5: Continuously improve reseller inventory levels (Marketing Manager)				4							-30			

Plan Tag: Revenue and Profits **Objective/Strategy Tag:** Profits
Report produced by PlanBase software. ©1998-2011 PlanBase Inc.

Figure 10-12: Developing an Implementation Plan from a Hoshin Plan Strategy.
Each strategy in a Hoshin Plan must have a supporting Implementation Plan, unless the strategy is to be deployed further. Here we show the Implementation Plan for the Marketing Manager's strategy number 1.2.3 from her Hoshin Plan in Fig.10-5. This format, provided courtesy of Plan Base Inc., was developed via their web-based Hoshin Planning system. It shows the traceability up to the General Manager's Objective 1.0 (from Fig. 10-5). This specific plan also records and communicates the review done to track progress of the implementation plan.

Daily Management Plan

The Daily Management plan focuses on keeping the house in order; that is maintaining the performance of day-to-day or repetitive processes. No special effort other than establishing goals, control limits, and a monitoring system are required. The perquisite is that these processes are well understood because there is a wealth of experience and knowledge.

The Daily Management plan is used to monitor, control, and manage critical day-to-day processes. If there is a deviation from target in the measured process it must be resolved quickly to prevent recurrence. The recommended format is shown in Fig. 10-13 and a completed plan is shown in Fig. 10-14.

Elements in a Daily Management Plan:

- *Item:* List the item to be managed. This plan has items that are well understood and documented. Little improvement is required for this item and the expectation is that customers and management are happy with the target and performance of this item.
- *Goal and control limit:* Each item will have a goal, and all goals must have control limits. When a process deviates outside of its control limits, the deviation must be analyzed, understood, and recurrence must be prevented.
- *When reviewed:* Here we list when the item is reviewed and checked.
- *Data source:* Here we list the source of the data for the item under control.
- *Owner:* Ownership or responsibility must be clearly defined.

Guidelines for Daily Management Plan

How many items should there be in a Daily Management Plan? *We recommend an average of ten items for each manger or department.* This is a guideline: the concern we have is that it may be difficult to manage too many activities. Hence priorities must be set. Some examples of potential items are given below for a manufacturing operation.

- The daily production plan for: Targets and performance.
- Productivity: Actual data Vs targets
- Production yields and other wastes.
- Customer satisfaction index, via customer complaint resolution, product usage satisfaction, on-time delivery.
- Product quality and reliability.
- Kaizen or improvement activity
- Overall cost tracking, including actual expenses Vs targets.
- Safety and employee morale.

DAILY MANAGEMENT PLAN

No.	Item	Goal & Action Limits	When Reviewed	Data Source	Actual Performance Jan.	Feb.	Mar.	----	Owner

Figure 10-13: Format of Business Fundamentals Plan

DAILY MANAGEMENT PLAN

Department: Operations | Manager: Operations Manager | FY: 2011

NO.	ITEM	Goal & Action Limits	Review Date	Data Source	ACTUAL PERFORMANCE JAN	FEB	MAR	--
1	Customer complaints resolution	Short Term Resolution in 24 hrs. Limit: 36hrs	Monthly	Sales & Service				
2	Product annual failure rate (All product average)	0.5% Limit 0.7%	Monthly	Service				
3	Delivery Per Commitment	100% Limit 97%	Daily	Customer Report				
4	Incoming Inspection of Critical Parts	< 500 ppm	Monthly	QA Dept				
5	Production loss, all lines (corollary of yield)	< 0.5 %	Weekly	Eng. Dept.				
6	Outgoing: Out of Box Inspection	< 0.1 %	Daily	QA Dept				
7	Expenses at target	100%, +0%, -10%	Monthly	Admin.				
9	Cost Reduction Program	10% /Quarter Limit 8%	Quarterly	Admin.				
10	Employee Atrition Rate	< 12%/Annum	Monthly	HR				

The figure shows the Operations Manager's Daily Management Plan. Note that this lists routine items that are done on a daily basis. These are items where there is a wealth of experience and knowledge - all these items should have standards on how they are to be performed. Hence, the goals and limits are based on experience and no implementation plans are required.

Figure 10-14: Daily Management Plan

Displaying Daily Management plan

It is a good practice to display Daily Management plans (or KPIs) at each department or production line. In such cases the relevant metrics should be posted on the production line so that operators, supervisors, and engineers get real-time information on their daily performance. This way all staff can see daily progress, issues, and countermeasures taken.

Managing Abnormalities and Deviations

Management of Daily Management plan requires that that we discover abnormalities and deviations from target and we analyze and prevent their recurrence. This is important and a system must be in place to allow for it. Good documentation helps. There are several methods for doing this:

Use a Hoshin plan review table: This is discussed and illustrated in the section entitled "Review of Hoshin or Daily Management Plans". A review table is useful if deviations are reviewed regularly.

Use the standard Daily Management form: The form can be modified with additional columns to explain the deviation and the proposed countermeasures.

Use an out-of-control report or CAR (corrective action report or request). This is a standard industry format for analysis in PDCA format.

Setting Numerical Targets

The Numbers Game: Numerical targets have to be set for both objectives and performance measures. When this is done, the numbers must be derived with the desire to have a breakthrough objective (for a Hoshin objective) or data (for Daily Management Plans). The following are guidelines:

- What was the achievement over the last few years? What has been the range of performance?
- For improvements and breakthroughs, a rule of thumb is 50% improvement.
- What have major competitors achieved?
- What are the trends in the industry?
- Do we wish to exceed past performance and by how much?
- What do our customers expect?
- What are the market expectations and opportunities?

The Review for Hoshin Plans and Daily Management Plans

Basic Methods of Analysis for Conducting a Review

The annual planning process is not complete without comparing progress of the Hoshin plan to the targets. Any deviation between the actual and intended results must be analyzed in order to determine the reasons for the deviation. If there is deviation, we need to take countermeasures.

Even if the results are satisfactory – that is performance is very good – the results must be reviewed. This means determining the appropriateness of strategies and performance measures that were established during the planning process. It is important to understand the reason for success and failures: Only then can learn to do better and use the learning to shape future behavior and plans.

The review format is shown in Fig. 10-15. An explanation of each element follows. A completed example is shown in Fig 10-16. Note that the review format is segmented into four elements which represent the PDCA cycle.

- *Objectives and strategies:* Each objective and its strategies from the Hoshin plan are listed and summarized here.
- *Actual Performance:* Actual results are listed for comparison with goals (of the objective) and performance measures (for strategies).
- *Status Flag:* We recommend using flags as they provide a quick, visual, summary of performance. The flags also provide quick visual management. An explanation of the flags:
  - *Green* or on-track: Used when a goal or performance has been met. However if the goal is to be met at a later date (after the current review), then this indicates that we are proceeding according to implementation plan.
  - *Yellow* or warning: Used if there is a strong possibility that a target of performance measure will not be achieved, or if the plan is not progressing well. Management intervention is required.
  - *Red* or off-track: Used when failure to meet plan is about to happen. Management intervention is required.
  - *P* or inappropriate process, strategy, or performance measure: Used when we discover that our strategy or measure is inappropriate. This often requires a mid-course change during the year.
- *Analysis:* Here we state the cause for the difference between the Plan and Do stage. When there is deviation, it is important to understand the root cause – ask why five times!
- *Implications:* The outcome of this phase will influence the direction or any changes in the plan between now and the next review.

HOSHIN PLAN REVIEW TABLE					
Prepared By:	Date:		Year		Location:
PLAN: Objective & Strategy	**DO:** Actual Performance	Status Flag	**CHECK:** Summary of Analysis and Deviations		**ACT:** Implications for Future
		G,Y,R,P			

Figure 10-15: Hoshin plan review table

Guidelines for Conducting a Review

Here are some hints for conducting an effective review:

Frequency: For any organization or operation, we recommend monthly or quarterly reviews of the Hoshin plan. However, at the department level, the implementation plan should be reviewed more frequently to ensure progress is being made. Similarly the Daily Management plan should be reviewed daily or weekly (within each department) to ensure all items are in control.

Format: The recommend format for the Hoshin plan review is shown in Fig. 10-15; a completed review is shown in Fig. 10-16. When reviewing the Daily Management plan, it is advisable to add comments in the right-hand margin.

Review procedure: For the Hoshin plan a formal review should be done, with the management team in attendance. Here are some presentation guidelines:

- Discuss objectives and strategies.
- Review goals and performance measures.
- If there is a deviation, state the reason for the deviation in the analysis column and the countermeasures in the implications column.
- If progressing to target, put comments and learning points in the implication column. The review table is in PDCA format, and the cycle must be completed.
- The manager under review should be concise: List the facts, root causes, and countermeasures concisely. However, be prepared with back-up explanation data, since the review format has room only for a summary.
- If a target is scheduled for the fourth quarter, it will be insufficient to say you are on track early in the year. The danger signals that may show are:
  - Comment during the first, second, or third quarters: OK, on-track.
  - Comments during the fourth quarter: Oops! Missed goal, did not have the resources.

- - o Therefore you must check progress by reviewing the milestones in the implementation plan. If there is no implementation plan, there is a problem. If there is one, ensure it is reviewed monthly and the milestones are on track.
    - o Always be wary of comments such as: no problem. Request backup data and check for progress.
- For the fourth quarter or final review of the year:
    - o Use the same procedure to generate issues for next year's plan.
    - o Typically, the third quarter review often becomes the final annual review. For this we suggest the following: Review the actual performance up to the third quarter and provide a prediction for the balance of the year. The prediction must be based on data and trends. This will give you inputs for the next year's plan.

Duration of review: Each manager should take 30 to 60 minutes. Therefore an organization review should take up to one day for a full review.

Changing objectives or strategies during the year: Often an organization may discover that an objective is wrong or inappropriate. Other times there may be external factors requiring a change of direction. Making the change in the plan is therefore appropriate.

Hoshin plan reviews are done bottoms up: Hoshin reviews should be done bottoms up. That is, start with implementation plan reviews at the lowest levels, followed by the next level up Hoshin plans, and finally the review by the General Manager or CEO of the organization. This bottoms-up review process ensures that when the final high-level review is done, there is confidence that all plans have been reviewed and countermeasures have been taken to ensure progress.

Implementation plan reviews: Implementation plans must be reviewed regularly: Weekly or monthly reviews are best. Implementation plan reviews are essential for success; this is the point where all the best laid plans will culminate into results – we recommend extra attention for this activity. As the plan must have schedules and checkpoints, these must be checked for progress. Deviations must be understood and countermeasures taken. Comments or changes can be added on the implementation plan. Fig. 10-12 shows a screen-shot of an implementation plan review table, which combines both the implementation plan and its review. After the implementation plans are reviewed, the Hoshin plans are reviewed bottoms up.

Doing the Right Things - Hoshin Kanri Planning

HOSHIN PLAN REVIEW TABLE: 3Q2011					
Prepared By: Operation Manager	Date: : August 2011		Year 2011	Location: Prod. Div. Apex
(P) OBJECTIVE & STRATEGY	(D) ACTUAL PERFORMANCE	Status Flag	(C) SUMMARY OF ANALYSIS OF DEVIATION	(A) IMPLICATIONS FOR FUTURE
2.2 Reduce manufacturing costs 2.3 Reduce externeal product failure rate	Target 1: Cost reduction of 30 % Actual: Overall reduction of 16 % Target 2: Reduce Fail rate 4 to 1 % Actual: Currently at 2.5 %			
2.21 Review and reduce procurement costs for all products with life of more than 8 months.	Perf. Measure: Cost reduction for product A, B, C of 15% Actual: Achieved 19 %	Green	We achieved this through active negotiation and use of lower costs substitues parts: E.g. Plastic base instead of sheet-metal.	
2.22 Redesign and simplify product A and B. These products have life of 2-3 more years.	Perf. Measure: Reduction in manufacturing costs for product A, B of 15% Actua:16%	Green	Target achieved through extensive redesign. New design guidelines shared between R&D and Operations.	
2.23 Increase outsourcing of non-critical sub-assemblies by April.	Perf. Measure: Target for 50% increase in outsourcing. Expect 30% cost reduction. Actual: 55 % of sub-assemblies have been outsourced. Cost reduction is only 22%.	Red	Supplier is not able to achieve the cost reduction of 30%, but freight cost from new suppliers erased some benefits..	Assembly Cost reduction estimate was given by supply chain team: Assembly cost reduction of 35% was achieved. In future total supply chain costs must be computed.
2.31 Reduce failure rate of products B, C, & D.	Perf. Measure: Reduce top 5 failures by 50% Actual: Currrently at 28%	Yellow	All countermeasures and plans have been implemented. Results are not visible because it may take several months before the effect is seen in the field.	No action planned. Expect to reach 40 % by 4Q '11 and meet 50% by 2Q '11.

Figure 10-16: Hoshin Plan Review Table

Putting All the Plans Together

Let's collect our thoughts and put together the ideas we have discussed on the planning process. It begins with the company's vision, economic factors, and customer and market requirements. The process then moves on to preparing the Hoshin Plan, Daily Management plan, and the Implementation Plan; it finishes with regular reviews of progress.

We illustrate the cycle in Fig. 10-17, which summarizes the required activity. A very important step will be preparing the implementation plan. This is the point where all the best laid plans will culminate into results – we recommend extra attention for this activity.

Planning and Budgeting

The annual budget and sales targets must be tied closely to the planning process. The budget must support the planned activity; hence it is crucial to finalize the budget after the plans are completed to ensure resources match plans.

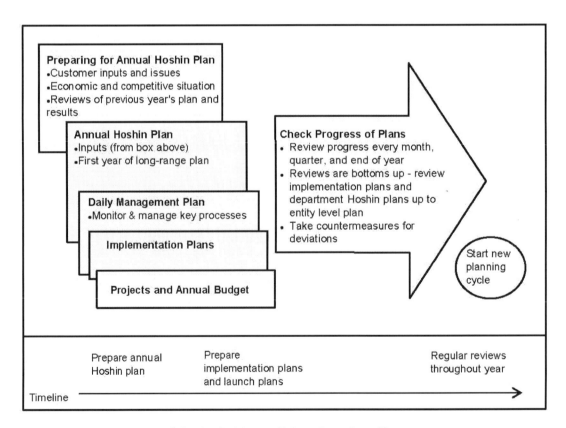

Figure 10-17: Putting all the plans together

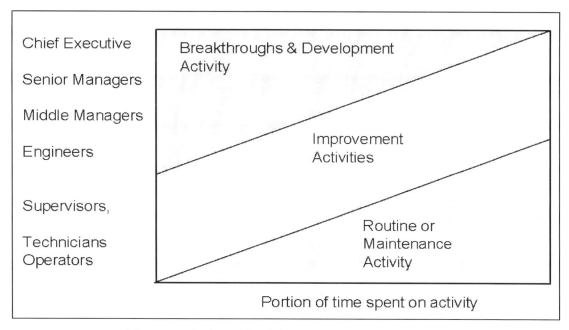

Figure 10-18: The Itoh Time Management Model

Planning and Time Management

One of the constant complaints of managers is that there is no time to implement planned activities: Often plans are completed and put away, and everyone goes back to tackling the activities at hand. Hence, we would like give the missing answer to the perennial question: *"How should we manage our time at work?"* For the answer to this question, look at the Itoh[42] time management model. This is shown in Fig. 10-18. You read it from left to right to see the layer of management or employees that are responsible for each activity.

Breakthrough or development activity: This is the primary activity for a successful organization. This includes development and introduction of new products and services; steering the organization in the correct direction; revenue generation activities; investments to grow the business; and develop and strengthen the organization. Clearly this activity is the work of senior managers. This chief executive of the organization probably only does this activity and rightly so. Many professionals, designers, and engineers would also be involved in this activity.

Improvement activity: Continuous improvement is another important activity. An organization cannot rely on a continuous stream of blockbuster products and services. Many of the current processes, products and services will require improvement in efficiency, productivity, and quality. Senior management's role is to identify, prioritize and review progress in this area. The bulk of the work will be done by middle managers and professionals, such as engineers and accountants.

Routine or maintenance activity: This refers to management of routine day-to-day processes. These processes require to be maintained at their current level of performance. This would include things like running production in an automobile, computer, or turbine engine factory. These activities represent the fundamental processes of an organization, with processes which are running routinely and require little management intervention, so long as they are in control. This activity is done primarily by the operators, supervisors, engineers, and professionals.

Comments and Criticism of Hoshin Kanri Planning

Hoshin Kanri has been criticized for being too detailed: According to William Dettmer[43]: *"It luxuriates in the details of record keeping, data collection, calendars, schedules, frequent and rigorous plan and progress reviews, and the roles and responsibilities associated with these activities."*

This is a valid criticism and there is a danger that this planning process ends up creating a burdensome bureaucracy that stifles creativity. Some ways to avoid this have been discussed:

- The highest level Hoshin Plan in the organization should focus on one breakthrough activity or objective.

- There should be one (best) or two (maximum) layers of plans supporting the CEO's plan; beyond that there should be implementation plans or project plans only.

- The number of documents and plans in the system can be minimized by using the alternate Hoshin Plan deployment matrix in Fig. 10-11. You can also use web-based electronic formats to minimize paperwork. However there is still a need for implementation plans and regular performance reviews.

- If the process is managed well, Hoshin Kanri is a very complete planning process that goes beyond the traditional approach of strategic planning. Even its most vocal critic, William Dettmer, agrees: *"It promotes 'doing' and 'review' as well as planning. It incorporates operational and project planning, as well as strategy formation."*[44]

- The next section discusses planning using TOC (theory of constraints) methodology. Per the discussion and lesson in the TOC session, we propose only ONE goal for the organization. This specific application from the Delta Institute is often called *Super Hoshin planning* as it provides more focus.

The Theory of Constraints

The theory of constraints (TOC) was expounded by the late Dr. Eli Goldratt. He popularized his theory in a business book, actually a fast paced novel: "The Goal". In his book he discusses the concept of the system constraints that hinder progress and profit in a business. The rest of this section on TOC is an edited extract (we are responsible for any errors or omissions) from a write-up by Dr. Dieter Legat[45] of the Delta Institute.

The theory behind "The Goal" comes from German Manufacturing Theory of managing Takt time or *taktzeit,* which strives to continuously remove the weakest link or slowest time in a series of manufacturing steps, in order to remove bottlenecks and speed production; this is an ongoing cycle and as each weak link is removed the operation gets more productive and efficient. TOC theory requires you to think as follows:

View the business as a system: Goldratt, however, expands this principle of removing the weakest link to the business organization. TOC theory takes a wide view of businesses: They are viewed as systems rather than a set or processes. TOC posits that the business system consists of company processes, policies, and the entire eco-system; the eco-system includes customers, distribution channels, government policy, suppliers, and competitors.

Set the right goal: According to Goldratt, business leaders often fail to focus on the right goal. This is especially true if a company is in trouble; at that point management will focus on cutting cost rather than on increasing "throughput", which is equivalent to net contribution margin (revenue – variable cost). If we focus on throughput as the prime goal of the company then problem solving requires a different frame of mind compared to cost cutting, which often means downsizing.

Find the constraint and resolve it: Here we use TOC planning to resolve.

Using TOC for Business Planning

In a business there are normally two situations; let's review both situations and review how TOC can be used to manage the business:

Situation A: The business is going well; in this case management must create a plan to continue growing profitably. Therefore management must *predict* future constraints to meeting its goal and resolve them. The TOC tool for developing the resolution plan is called "prerequisite tree" – which is a cause & effect analysis done "backwards from the future". If the constraint is predicted correctly and the solution plan is effective, the company will reach its goal.

Situation B: The business is in trouble; in this case management must create a plan to get back to growing profitably. Therefore management must find the constraint and resolve it. If the constraint is correctly resolved, sales will turn back to profitable growth.

Therefore in both situations A or B, the next step in TOC planning is: *Establish the goal*: This should be the prime goal for the entire company. So what is the purpose of a company? According to the late David Packard it is *"to add significant value to the customer, which will become visible by the company's revenue minus the company cost"*. Thus the goal of the company is to have throughput and make money. This is the only goal of a company. In a non-profit organization the goal will be different, but in our context we are addressing a for-profit business and its manufacturing operation.

For situation A, after setting the goal, the next step for the business that is doing well: *Determine the critical success factors and necessary conditions*: Critical success factors are conditions which the system must achieve in order to reach the goal (like customer satisfaction or product quality). Necessary conditions are the measurable attributes of these critical factors (like "ranked in top 3 vendor list" for customer satisfaction).

Determine the obstacles that prevent the business from reaching the *necessary conditions*. The obstacle to a critical success factor of customer satisfaction could be poor customer service (in the area of face-to-face interaction or product repair) or excessive product failure, or unreliable automobiles. In this phase, a "logic tree" of cause and effect is often used to determine the root cause; this is similar in concept to the cause and effect (Ishikawa) diagram.

For situation B, after setting the goal, the next step for the business that is not doing well: *Determine the constraint*: What is the problem? We need to find the constraint and its root cause: In TOC terms this root cause must be searched for by rigorous cause and effect analysis through the system to ensure that we actually find the true root cause. Next, d*etermine the obstacles* that prevent us from eliminating the root cause. Only then can we have a plan for overcoming the obstacles.

Next, for both situations define the plan, execute, and review progress: *Define the plan to overcome the obstacles.* Here we determine the tactics required to overcome the obstacles (required result, tactics owner, budgets, and schedule for completion). This is similar to the implementation plan in Hoshin planning.

Execute the plan. The strategy owners execute the plan. Constraint management theory states that we must *"subordinate everything else"* in the organization and work only on executing this plan. To subordinate everything else implies that we should not work on other major issues – this is the same concept that we would apply for cycle time improvement: focus on one assembly step at a time.

Review progress and fine-tune the strategy based on progress per the planned milestones and towards meeting the goal.

Achieve the goal and then start the cycle again by planning to work on the next weak link or constraint in the business.

Goldratt recommends a cyclical approach (like the PDCA cycle). Meaning: after resolving the constraint the next one will appear somewhere else, and we start the cycle again. Just like a production line, we strive to continuously remove the weakest link in a series of manufacturing steps, in order to remove constraints and improve efficiency. The above steps fit nicely into a PDCA cycle format. This looks a lot like Hoshin planning. But it does have one major difference: It emphasizes only ONE goal, while Hoshin planning allows for more than one.

Refer to Fig. 10-19: Here we show a business plan using TOC methodology. The Company has one goal which is revenue of $120M. The necessary conditions are given as breakthroughs in key accounts, strong growth in base business, and product development. All of these conditions have targets. The obstacles and the deliverables to overcome the obstacles are discussed and documented. These strategies have owners, timelines, and the budget for achievement.

Typically, in this phase of planning, a "logic tree" of cause and effect is often used to determine the root cause of the obstacles. The typical process of developing a TOC business plan requires "logical trees". These are an excellent tool for analysis, but may fail in providing a good plan. Note: A sound logic tree may easily span 2-3 flipchart pages – this is just too impractical to use in day-to-day management work. We recommend that you combine the format of Hoshin based operational plans with the power of TOC logic trees into one single template in the "the one page operational plan", which is shown in Fig 10-19. This approach of ONE goal with a focused effort to achieve the goal can be called a "Super Hoshin" plan.

Constraint management is a successful methodology and allows focus and quick execution of improvements towards the organization's goal. Goldratt's theory of coupling a Goal with TOC and constraint management at a system level is brilliant; his focus on throughput (product or sales) instead of cost, customer satisfaction, or inventory, is right-on. In fact improving throughput will help to improve customer satisfaction and bring down costs and inventory, but still they must be tackled when other constraints are removed.

We have given a quick but powerful summary of constraint management. More details on TOC and references can be found in the appendices. Alternatively, you can take the lessons from the TOC constraint management process and use them to improve your Hoshin planning.

Operational Plan

DELTA INSTITUTE SWITZERLAND

Goals			Plan					Progress
Overall Goal	Necessary Conditions		Reasoning (Obstacles)	Commitments				J F M A M J J O N D
	Focus	State		Deliverable	Owner	Due	Budget	
120 M$ in 2008	1. Breakthrough in Key Accounts	T-Funnel of 200 M$	Account constraints are unknown	ACA for all key accounts done	FB	July		
			OppMap is not reviewed with accounts	OpMaps reviewed at account meetings	KS	Feb		
		>50% Win Rate in T-Projects	Win management is weak	All projects > 10M have managed win plans	KS	Jan		
				W/L reviews for all projects >10M	AF	March		
	2. Strong growth in base business	C-Reach > 65%	C-Reach goes untracked	C-reach reviewed at monthly management meetings	NE	Jan	125 k$ CRM	
			Ownership is unclear	Sales defines and selects owner	BR	Jan		
		>10% CR	Ownership and goal are undefined	CR reviewed at monthly management meetings	DL	Jan	50 k$ CRM	
			Improvement plan is not defined, confirmed	Improvement plan reviewed at monthly management meetings	DL	Jan		
	3. Fast Product Development	< 6 months for new products	Goal is not key company performance standard	PD Cycle of 6 months is communicated as one of "top ten goals"	BR	Jan		
			Project methodology is unknown	Project progress reviewed weekly (Friday afternoon)	BR	March	200k$ Consult	

Fig 10-19: A business plan developed with the TOC process. This format uses the "the one page operational plan" developed by the Delta Institute. Note that this template contains both the Plan and Do steps of the PDCA cycle. Reproduced with permission.

Using the Lessons Learnt from TOC in Hoshin planning

Both Hoshin and TOC planning have similar aims: to resolve problems either proactively or reactively. They both have strengths and weaknesses; we believe that merging the strengths of the two is the best way forward. Some of the good points of TOC can be incorporated into Hoshin planning; the result will be a more effective and focused plan. Let's discuss how we can do that.

Focus your efforts with ONE goal; recognize that this goal is centered on increasing revenue or profits for the business. Everything is subordinate.

Recognize that Customer Satisfaction or Quality is not a business planning goal, but it is a critical success factor or necessary condition to achieving higher revenue and profits. Therefore any kaizen projects must support the critical success factors.

Determine the necessary conditions for increasing revenue and how to overcome those obstacles; refer to the steps in the TOC example.

Reduce complexity by using the one-page format used in Fig. 10-19. Consequently the Hoshin plans we shared with you earlier in Figs. 10-5 and 10-6, can be merged under one goal: Revenue at X % profit. You will still need implementation plans and regular reviews to ensure good progress.

Leadership in the Company or Organization

There are many managers in a company but what we want is leaders. What's the difference? Per our quote at the beginning of this chapter: *The manager does things right; the leader does the right things.*

We feel leaders can move an organization towards success and the right position in the marketplace. There are, however, some good traits that a leader must have and if these are well executed during, they can help drive the organization to success. Some of the traits are:

Create a vision: When we discussed the Hoshin planning process, we mentioned the need for a company or organization vision. Leaders create a vision for the future, often by articulating a dream. Leaders understand the need for a vision to harness, motivate, and guide the organization. Leaders also understand that a vision is more than just a goal in the future. A great vision incorporates values and a long term goal. Dr. Deming[46] in his 14 points calls this: *Create constancy of purpose* for the improvement of product and service.

Set a clear goal: The vision will have both value and a long term goal where the company is going. An aggressive goal will provide new challenges and excitement but will take many years to implement; hence, the strong leader must be consistent, unwavering, and continue to steer the organization through thick and thin towards the vision. This will require leaders to *"create constancy of purpose"*. For example, NASA's goal was to reach the moon. Closer to home, Toyota's vision is "Be number one in quality and offer good products to respond to requests of society and trust of customers." Its previous goal set in the early 1980's was "Global 10" – a goal to gain 10% market share worldwide by 1999. That goal was achieved.

Set priorities and then focus the organization to achieve them. Leaders understand that there is more to do besides working towards achieving the vision. So, priorities must be set, after which the organization must focus its energy in achieving them. If this course of action is followed, results are achieved, employees are motivated, and the cycle can be repeated. In line with this, we quote Dr. Deming again in one of his 14 points: *Improve constantly and forever the system of production, service, planning, or any activity.* This will improve quality and productivity and thus constantly decrease costs.

Stay in touch with customers and meet their needs. Leaders must continually meet their customers regularly and pulse them for their needs. They must strive not only to meet those needs but drive the organization to exceed them. Meeting customer needs includes providing basic functions such as impeccable product quality while exceeding needs includes providing the attractive, exciting, and unexpected.

Empower the organization. An effective leader does not micromanage because there is always much to do. The leader delegates and coaches and encourages a strong management team to emerge and grow in order to implement the necessary strategies that will lead to success. This includes hiring strong people, training and educating employees, giving tough assignments, and delegating strategies to lower level managers.

There is always a shortage of leaders in companies, organizations, and nations. We have no magic formula for leadership but have listed some basic traits of a leader. A leader who practices these traits will certainly be able to lead the organization down a successful path.

Summary and Conclusion

Planning is one of the most important processes in an organization. It is what drives most other activity. We have proposed the use of the Hoshin planning methodology. This requires an annual Hoshin plan for identifying and managing breakthroughs, with aggressive goals in an organization; plus a Daily Management plan for controlling and managing critical day-to-day processes.

During planning we recommend that you focus on reviewing four critical areas: Quality, Costs, Delivery (of current and new products), and Education. This is to ensure you do not neglect these important, generic, success factors.

The Hoshin planning methodology is a systematic and tightly coupled planning process. It requires effort and consensus and in return provides a focus and a single-minded approach by the entire management team to move the organization forward. Hoshin planning provides many benefits, including systematic thinking, better coordination, sharper objectives, improved performance standards, and management involvement. All this results in a planned approach to tackling the marketplace that eventually can end in higher sales and profits.

We briefly discussed TOC – theory of constraints – and discussed how some of its concepts can improve Hoshin planning.

We have discussed most of the methodologies that help foster operational excellence. In the next chapter we review Value Stream Mapping and discuss how it can be used to identify opportunities for improvements in a company; after that we will wrap up and discuss how we can use all the methodologies to achieve operational excellence.

Chapter 11

Value Stream Mapping

That is the exploration that awaits you! Not mapping stars and studying nebula, but charting the unknown possibilities......
Leonard Nimoy

Overview

In their ground breaking text *"Lean Thinking"*, Womack and Jones discuss the value stream. The *value stream*[47] is the set of specific actions that are required to bring a specific product (goods or services) through the three critical management tasks of any business: The *problem-solving task* from concept to production, the *information management task* from order to delivery, and the *physical transformation task* from raw materials to finished product. Once this is understood, the next step is to identify the entire value stream for the product and identify waste, or *muda*, which must be eliminated.

In this chapter we will review:
- Value Stream Mapping.
- Value Stream Mapping procedures in manufacturing.
- Value Stream Mapping projects.
- How to identify opportunities to achieve the "perfect" future state.

Value Stream Mapping

Value stream mapping (VSM) can help us map the current state of our design, information management, or physical transformation activities. Once mapping of the current state is completed, an enormous amount of waste can be exposed and improvements can be planned; this will lead us to a future state of "perfection".

Womack and Jones propose that we to identify the *entire value stream* for the product, but this is a step that most firms rarely attempt. In our discussion we will discuss VSM and improvements for the order to delivery process, and the physical (product) transformation task.

When we achieve an improved future or "perfect" state we create value for our business and for the customer. *Womack and Jones have emphasized that creating value is the first principle of lean thinking.* This is the ultimate aim of value stream mapping: Create Value.

VSM is a practical tool and has its place in the hierarchy of tools we have discussed so far. VSM has the ability to put all necessary information of the current state of a firm or operation into a visual form for management to see and analyze. The opportunities for improvement can then be identified and the future state and action plans can be developed. The current state will surface issues such as pockets of high inventory, production delays and bottlenecks, excessive transport time, high defects, and high cost areas. But it is only the starting point of the improvement journey.

Value Stream Mapping Procedure

To get started on VSM, a project team needs to be formed and a leader appointed. VSM projects can be for one operation, cover several operations, or span an entire company. However, for a small production line it may be more productive to use the *gemba walk* approach and draw a process map as we did in Chapter 8. The basic VSM steps are:

Appoint a VSM project team. The team members must be familiar with the product and understand the process flow from supplier to manufacturing to the customer. For a multi-operation VSM that spans across a company, representatives from each of the various operations must be pulled in.

Select the operation or product for the project. In the chapter on Quality, we gave suggestions for project identification and selection.

Go to the shop floor to get the facts and data: It is crucial you get the facts and data directly from the production floor; for larger operations, involving several sites, a visit to all sites is essential. This is similar to the *gemba w*alk, which we discussed earlier. For a multi-operational team, data and information has to be pulled in from the various operations.

Draw the VSM map. This will represent the product and process flow from supplier to customer. The map must show both physical and information flow. For very large groups some detail must be sacrificed as in Fig 11-1. Here is a list of required information to get you started:

- Incoming inventory and supplier transit times.
- Incoming suppliers and incoming quality inspection.
- Each workstation: defects, cycle time, inventory, one-piece Vs batch.
- Each test station: defects, cycle time, inventory, one-piece Vs batch.
- Each inspection station: defects, cycle time, inventory, one-piece Vs batch.
- Equipment, workstation, and product changeover times.
- Storage locations and quantity between work or test stations.
- Finished goods inventory at the factory and other locations

- Finished goods inventory at distribution centers.
- Number of operators at stations and warehouses.
- Floor space at each location.
- Transit times to distribution centers and customers.
- Information flow for orders, response times, lead times. This is especially useful when you draw a VSM for multiple locations and geographies.
- After all workstations are sketched into the VSM map, draw a cycle time graph below the stations to show cycle times, changeover times, and non-value added time. This will help to highlight opportunities. Refer to Figs 11-1 and 11-2.

Use the appropriate symbols in your VSM map and record the information in data boxes as shown in Figs. 11-1 and 11-2. Additional information is available from the references in the appendices.

Double-check and categorize the important points at each station or location, plus complete the VSM map: This includes inventory levels, lead times, defect levels, people count, cycle times, changeover times, and floor space. These categories represent cost and effort from which waste can be identified. Finalize the map, share with the team, and discuss its accuracy.

Review and analyze the data in the VSM map. On a well-drawn map, waste can be easily identified. Some obvious opportunities may be apparent but an understanding and intense review of current practices and obstacles may be required before you come with new ideas and priorities. Look at the map and review:

- The major process areas: cycle time, changeover time, resources, capacity.
- Bottlenecks in the system, due to changeover or capacity.
- Low (good) and high (bad) inventory areas in the system.
- Other waste in the system: defect rates, excess transit times, non-value added time, defects, excess people, excess floor space.
- How information flow (electronic and manual) impacts physical flow.

Ask the question: *Does this* activity add value to the customer? *Repeat: Value to the customer – not to the company.* If the answer is no, you have identified a potential opportunity for improvement.

Draw up a draft future state map. From the analysis and waste identification in the previous step – you will be able to identify potential areas for improvement. At this point you should review the current purpose of all processes in the VSM map; and then discuss potential obstacles of removing any waste or bottleneck. Sketch out the potential future state below the current state and write down the target parameters for cycle time, changeover time, inventory, floor space, and so on.

This is the starting point of the improvement process. You will need to identify resource issues, process changes, training, timelines, and critical areas to improve. Set priorities. Finally, get approval to proceed if required.

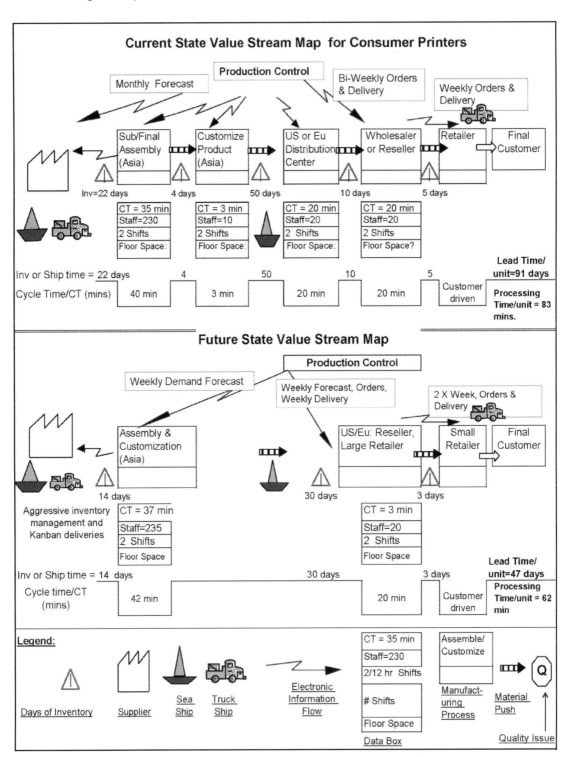

Fig 11-1: Value Stream Mapping for a printer manufacturer from supplier to end customer

Value Stream Mapping Project: Order Fulfillment

Let's discuss and review a value stream mapping project for order fulfillment, spanning purchasing, production, orders, and delivery. This is a high-level VSM project that spans a consumer printer business operating between Asia and USA or Europe.

In Fig 11-1, we show a VSM map for printer manufacturing from supplier to end customer. This is from our manufacturing experience and was prepared several years ago to look at ways to reduce lead times and costs. The value chain stretched from manufacturing in Asia, customization in Asia or (often) in US or Europe, stocking within the company at its own distribution centers in the US or Europe, sales to Wholesalers/Resellers, to retail outlets, and finally to the end-customer.

No surprise, that the actual travel time for a printer from raw parts to customer purchase was over 100 days. The customer can pick up her order in a few minutes, while the retail shop may have a lead time of 5 days for its order from the manufacturer, but the printer took almost a 100 days to travel from the factory to the customer. Clearly, mountains of waste!

Looking at entire current state, which existed several years ago, it is obvious that there were many areas that provide no value to the business or customer. These improvements are reflected in the future state map. Specifically:

- Merging of production and customization production lines at the factory.
- Inventory reduction of over 50 days.
- Elimination of massive warehouses in the US and Europe.
- Use of weekly sales forecast information to drive product customization.
- Direct weekly shipments from factory to resellers or large retailers.
- People count reduction in the factory and warehouse.

These are impressive improvements. However, the improvement process took several years. Why the delay? There were several reasons: High profitability, complacency, the improvement proposals needed top management approval, and it took time to implement.

The VSM done in Fig-11-1 is a good example of a high level map. From this map, major opportunities can be identified. Once an area is selected for improvement, for example merge manufacturing and customization, the team can do a detailed VSM map or process map at the next level and work on the next steps.

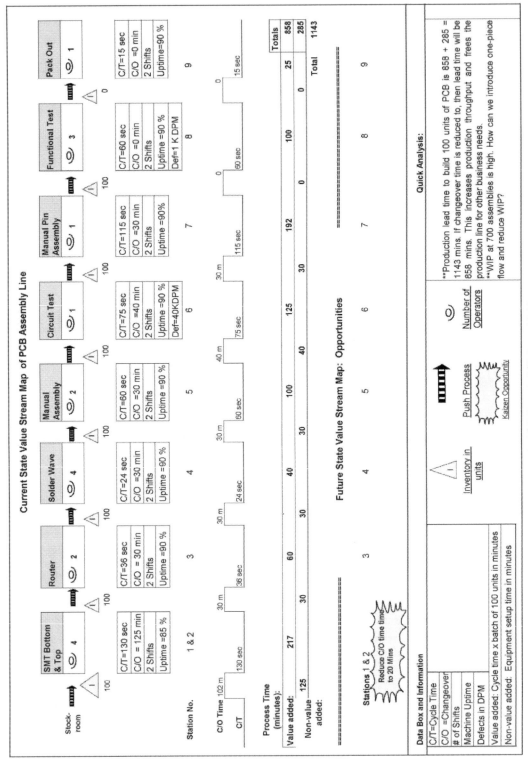

Fig. 11-2: VSM Map of a PCB Assembly Line.

Value Stream Mapping Project: Product Manufacturing

This second VSM project is for a large operation that manufactures assemblies and routers for Internet servers.

In Fig. 11-2, we show a VSM map for a production line manufacturing printed circuit boards, assembling, testing, packaging, and shipping. The line consists of SMT (surface mount technology) equipment, router, solder-wave machine, two manual assembly operations, an integrated circuit tester, a functional tester, and a pack-out station. The data boxes at each station show the cycle times, changeover times, uptime, and defect rate. The WIP and number of operators per station is also listed.

In the current state, it takes 1143 minutes to manufacture 100 assemblies. The analysis shows that the value added-time for manufacturing is 858 minutes or 76% of total time. The rest, or 285 minutes, is non-value added time. There are many opportunities to reduce waste or non-value added time. We have marked an opportunity at station 1 and 2, which is to reduce changeover time.

Typically opportunities are marked up within the current state map, after which the future state map is sketched out. We had a team consisting of engineers, production, and customers on this VSM improvement project. Why was the customer involved? Improvements will result in shorter lead times; this will result in a more responsive and cost-effective operation that responds quickly to smaller, or surprise, customer orders. What's more, customer engagement is a great thing to do.

We leave you to analyze and propose other opportunities. Based on what we have discussed so far, the opportunities include:

- Reduce changeover time reduction at SMT, router, solder wave, and testers. We have marked an opportunity at station 1 and 2, which is to reduce changeover time from 125 to 20 minutes, based on our experience in the changeover project in Chapter 6. This represent a clear opportunity.
- Reduce changeover time at manual pin assembly station.
- One-piece flow at workstations by eliminating batch flow.
- Reduce inventory via kanban delivery of material at workstations.
- Line balancing at the manual assembly and pack out lines.
- Review defect rates; The defect rate at the Functional Test is 1k DPM (0.1%) and looks reasonable. However, at Circuit Test the defect rate is running at 40k DPM (4%), which is high. Engineering will have to review the previous processes and reduce the failure rate.

These are potential opportunities; not all can be accomplished, but a review and analysis will present a short list of priorities.

Comments and Weakness of Value Stream Mapping

One weakness of VSM is that it's basically a waste identification technique. It does not place enough focus on product quality and customer issues. Another weakness is that it can be time-consuming and bureaucratic. Toyota and many teachers of TPS do not use this tool. They prefer the g*enchi genbutsu* approach.

Let's look at what they would do if they used the genchi genbutsu approach on the production floor in Fig. 11-2: A detailed tour with engineers and production staff would highlight the following: Review takt time and evaluate how it is impacted by long setup time, high inventory, batch processes, and defects at test. They would then set short term (3-Month) productivity improvement goals: Reduce setup time (by, say, 50%), introduce kanban at specific stations, reduce defects (by 50%) at test; appoint team leaders to review progress monthly. Then, return in 3 months and agree on the next improvement plan. To make this approach effective, there must be sensei, or experienced staff, to help conduct the audit. In contrast, the VSM approach would require a lengthy detailed analysis but a bright engineer would be able to come up with a useful map. Two different approaches, but both work and have their strengths.

Nevertheless, VSM can be very useful and has an important place in our tool box. The VSM order fulfillment project, in Fig. 11-1, clearly shows its benefits.

Summary

The VSM approach is especially valuable in looking at the big picture of waste and finding opportunity in a company or operation. However, it may not be suitable for every company or manager – because an overall value stream map of a large company will seldom have one owner, unless it is the President or VP. This was the case in one of our projects. But it is an excellent procedure to identify and remove waste in a manufacturing operation. The VSM process is beneficial because it can measure and document the waste in an operation, identify opportunities for improvement, and target for an improved future state.

Do realize that the VSM practice of a future state or "perfection" is the ideal situation, and it may be tough getting there; but it creates a good vision and direction. Remember, today's perfect state may be tomorrow's waste. Continuous improvement, TPS and Lean Manufacturing is a never ending journey. Nevertheless, as we move an operation towards the "perfect" state we create value for our business and for the customer. This is the ultimate aim of Value Stream Mapping: Create Value.

Chapter 12

Putting it All Together: Accomplish the Transformation

It is not necessary to change. Survival is not mandatory.
Put everybody in the company to work to accomplish the transformation.
Dr. Edwards Deming

Overview

By now you should be familiar with the different methodologies required to move towards operational excellence. You may be planning to start on this journey to transform your operation, or you may already have many of the recommended methodologies but want to improve and excel.

In this chapter we will discuss and review alternative approaches and challenges to accomplish the transformation to operational excellence. Specifically:

- Getting started: Some precursors and self-assessment.
- The Project Management approach.
- The Business Plan approach.
- Challenges and solutions for success.

Getting Started: Some Precursors

Understand Your Current Situation and Where You Are Going

The purpose of a business is to make profits and to have satisfied customers. Therefore any action plan to improve operational excellence must in the end achieve or enhance this purpose. All the tools we have discussed are the means to achieve this purpose.

Remember that Toyota came up with TPS because it had inferior products which could not compete with Western companies. Furthermore, some of Toyota's systems and processes were not planned for but evolved through necessity. Today, you have the luxury of reviewing and copying those parts of TPS that you can improve your productivity and competitiveness.

If you wish to implement the transformation towards operational excellence, including the basics of TPS and Lean Manufacturing, remember there is no end point or date. *It is going to be a never ending journey of continuous improvement.*

The project management approach and the business plan approach are based on our own personal experience; both models have worked for us. Which is the better way forward? The alternatives depend on the structure of your organization, its maturity, and the strength of the management team. Regardless of the approach, it is imperative that you educate and train your staff on why and how you plan to get started.

Why Drive Toward Operational Excellence?

First and foremost you need to communicate why you wish to start this journey; but if you have already started, you need to communicate where you are today and the next steps on the road ahead. There are many reasons why an organization wishes to move towards operational excellence. We list a few:

1. *An enlightened chief executive.* An enlightened chief executive will be aware of other companies' successes and realize operational excellence, based on TPS, TQM, and Lean Manufacturing, will lead to a better competitive position and higher profits.

2. *A company in a crisis situation.* In this case, the company management realizes that there is a crisis: products are not competitive, shipments are often late, the company cost-structure is too high, or there is a competitive threat.

3. *To increase profit.* There is much waste in every organization. Reducing waste will result in higher productivity, higher quality, lower costs, and higher customer satisfaction. The outcome will be higher profits.

Conduct an Overall Assessment

Our goal is to optimize and maximize factory performance. However, as we mentioned earlier: There is a tendency to disassemble the various tools into sub-sets which are promoted as standalone tools. For example: 5S, TQM, JIT, TPM, PDCA cycle, Six-Sigma, Lean Six-Sigma, Theory of Constraints, and A3 Process. This has resulted in a haphazard tool-driven attempt to copy these concepts and has delayed understanding and optimization of factory performance. Since our goal is to optimize and maximize factory performance, all the tools we have discussed must be used, but in the right circumstances.

So how do you start?

If you already have TPS or Lean program – look for opportunities to fine tune and excel. Hence, review current initiatives, strengths and weaknesses, and decide how to fill the gaps. An overall assessment to understand your current status will be useful.

If there is no aggressive program in place, the best method is to start by taking an overall look at the entire operation and understand strengths and weaknesses. Initially, focus on the whole and not any specific areas to improve. Start with a gemba walk of each operation or selected operations. Look, observe, and ask these questions:

- Is the workplace organized or cluttered? Is it clean or untidy?
- What type of waste can you see? Clogged aisles? Too much inventory in the warehouse or WIP on the production floor? Too much travel or motion by operators?
- What are the product yields? Is the production defect rate high? Are there boxes of product awaiting rework?
- Are there choke points and bottlenecks on the production line? Are the lines balanced or is there congestion and operator waiting time at workstations?
- How was the factory layout determined? Is the current layout equipment based or product based? Do you manufacture in cells or linear lines?
- Is equipment running smoothly? Does equipment breakdown often or does equipment produce defective parts due to poor maintenance?
- Is there a need to increase capacity? Can capacity be increased by reducing variability in the factory? For example variability due to batch production, lack of TPM, long machine setup times, high WIP, incoming vendor defects, unbalanced lines, excessive overtime? You must first look at variability before you consider adding machines and equipment.
- Are there robust metrics to measure factory performance? How are you performing to those metrics? Metrics reviewed should include: cost, delivery, quality, inventory, and productivity.
- Do you have a strong team of engineers, professionals, supervisors, technicians, and operators to manage the business?
- How is delivery to customers? Are you meeting commitments? Is there potential for improvement?
- Are there customer issues that need to be addressed? Are there many customer complaints? How serious are the customer complaints?
- An overall assessment to understand your current status will be useful.

Such a tour and list of questions will identify opportunities for improvement. This will help shape the direction, intensity, and priorities of your next steps.

Assessment Checklist

A detailed assessment checklist is provided in the appendices and will help you with a better understanding of your strengths and weaknesses; it will also help you decide where to focus your energy and next steps.

Educate and Train

The next step is to educate and train the manufacturing team in the tools. At this point, it's best to provide training for senior managers to discuss the key methodologies from this text. Once you move into specific projects, more detailed skills-training will be required.

The training material can come from a compilation of material from this text and other sources. Refer to the chapter on employee development and appendices for more details. The teaching must come from an experienced manager or leader from inside or outside the company. This teacher needs to be an experienced, patient, critical, and honest. A well meaning consultant with no practical experience, no management experience, or unwillingness to be critical will not be able to do this job.

Appoint a Facilitator or Sensei

Appoint a facilitator to help steer the transformation. The facilitator will help monitor and discuss issues and problems and help keep the team on track and schedule, plus arrange to get outside help when required. The facilitator or *sensei* – Japanese for teacher or master – should be a manager with experience in TPS, TQM, and Lean Manufacturing. The sensei's role includes:

Help in the overall assessment of your operation or factory

Challenge the team to select aggressive improvement targets, encourage good data collection, ensure no jumping to conclusions with a good decision making process.

Provide a deep understanding of the tools that will be used, and communicate the underlying principles of the tools.

Coach the team to do productive gemba walks to observe waste and identify opportunities for improvement.

Help the team understand the PDCA cycle and to use it effectively. This will enhance the learning process of the team. This is where the sensei's presence and contribution will be valuable.

However, it is important that the transformation must not be delegated to the sensei – his primary role is to train, advise, and challenge the team. To succeed the entire team must be involved, learning, and doing. With the precursors in place you can take the project management or business plan approach to accomplish the transformation.

The Project Management Approach

We have seen progressive general or operation managers start working independently on transforming their operations. Invariably, these managers get recognition and rewards, and become the leaders of change in the organization.

Select an Operation in Crisis or One Needing Improvement

You should select a production line or operation that is facing a crisis or needs improvement. For example an operation with high customer complaints, high defects, or high costs.

Appoint a Project Team and Lead

Once the area for improvement is selected, it's time to appoint a project team. Project team members should come primarily from within the team and must include both engineering representatives and production supervisors or leads. A project leader is also desired, typically an experienced manager in the operation who commands respect.

Revisit Safety Issues and the Organized Workplace

If your factory floor is well-organized and there is little need for improvement, move on. Otherwise it's best to take a step back and get the factory floor organized, using the 5S system. The 5S system lays the foundation for good manufacturing practices and creates an environment where problems can be easily identified and corrected. Furthermore, it helps to instill discipline in the workplace and it helps to set the stage for more advanced techniques. With this in place, you can proceed to the next step.

Work on the Most Critical Issue

Review the areas of customer or management concern. For example, there are numerous customer complaints and issues due to defects or late deliveries. After you have selected a critical issue analyze the problem, and prepare an action plan. In our experience, here are some potential crisis situations with solutions:

Defect rate too high. The defect rate experienced by customers is too high, or the internal defect rate is too high leading to rework and high WIP inventory. In this case quality needs to be improved. You would review the issues, get to root cause, and execute solutions; bring defects down to a decent level, where rework is manageable. Many of the possible solutions are discussed in Chapter 7: Quality tools such as Jidoka, poka-yoke, IPQA, and process controls. Remember if defects are high, it will be difficult to implement JIT (just in time manufacturing) and continuous flow; therefore it is crucial to keep defects low and processes stable before you can roll out other tools.

Late deliveries or missed shipments. You are not meeting the production plan. Potential problems impeding throughput could be:

- Defective products requiring rework.
- Defective material or delays in incoming material.

- Production line not meeting target production: The production line has bottlenecks and is not balanced, causing inventory buildup – refer to our discussions on cycle time, JIT manufacturing, and continuous flow.
- Machines giving insufficient output, creating bottlenecks.
- Batch production, which is unpredictable; hence it is impacting timely shipments.
- Variability of other processes in the operation that must be identified and resolved.

It is important to fix the most critical problems first using the appropriate methodology. Ensure you define a project size that can be completed in a short time frame of about three to six months. Results in a reasonable timeframe will encourage and motivate the teams and provide more management support.

Work on the next issue. Select the next project and continue with gains and success. This approach sets priorities and breaks the implementation into manageable and focused chunks of activity. Often you can work on several projects simultaneously – this is dependent to the project size and resource availability.

Value Stream Mapping. Value stream mapping (VSM) can be used in the project management approach. VSM can help you analyze the areas you wish to improve; after analysis you can use the appropriate methodology to improve. However, do not hunt for a critical issue to resolve in a company VSM map. Practitioners of the TOC approach would balk at such an approach; so would practitioners of the Hoshin planning approach

PDCA and Project Management

The project management approach fits neatly into the PDCA cycle, which gives a clear framework for managing projects.

- PLAN: Select the most critical issue; eliminate the root of problems, agree on the goal, prepare the detailed plan and schedule.
- DO: Implement the plan. Track and lead progress to the goal and commitments
- CHECK: Conduct regular reviews. Review the progress to plan. Identify any obstacles or wrong turns in the plan and resolve
- ACT: Continue to manage and monitor the plan; when it concludes, ensure the gains are institutionalized, documented, and in control. Look for the next critical issue to work on.

Our approach to problem selection has taken the following sequence: Review the organized workplace and safety issues, followed by quality improvement and productivity improvements. This sequence will have already helped reduce waste and cost. At this point it will be appropriate to look at other cost issues that need resolution.

In this step you can analyze areas of high cost and work on reducing them. This could be purchased parts, productivity, overtime, and so on; the analysis should look at the root cause of the cost issues and resolve them.

The project management approach works very well when you are resolving customer issues or operational problems. The biggest challenge with this project approach is that it is disconnected from the business plans, budget, or resources. Sometimes, if the project is very large, the project may be going in one direction, while the business may be going in another direction.

The project management approach may be considered as *extra work* by managers who are already stretched. Managers and other employees will give priority to managing the business and when they have time they will work on the improvement projects. Or they will just force through projects with a big stick: For example, we came across one top manager, whose operation was doing poorly in some of the tools we have discussed. On hearing our concern of slow progress, and wanting to please us and the CEO, he commented: "*...all right then, how many employee suggestions and kaizen events do you want me to achieve? Give me a target and I will get the operation to achieve it*".

Alas, this is a true and common story. It reflects on our initial failure to train this manager effectively and instill a cultural change in the organization, but it also reflects on a wider issue: the use of a carrot and stick to drive transformation. This is one of the reasons why lean/TPS programs fail. Therefore it is critical that the efforts are linked to real business needs and real customer driven projects.

Still, if the management team is committed, enthusiastic, and looking at customer and operational needs, the project management approach will work.

Table 11-1 (Summary of Figs. 10-5 and 10-6): General Managers' Objectives : Increase Revenue and Profits

No.	Strategy	Pertinent methodologies.	Timeline
1	Introduce new products and improve current products	• Review development cycle with the team, understand delays and obstacles and overcome them. Manage product development via the PDCA cycle.	<6 months, very urgent
2	Improve the supply chain	• Sales promotion and training of channel partners, review inventory levels and current processes; set new inventory targets.	<6 months
3	Develop system strategy for e-commerce	• Review and analyze e-commerce offerings, and propose and implement new systems to improve	<3-5 months
4	Revamp service business, increase throughput, and lower costs.	• Kaizen program to find better and faster ways to repair • Review repair standard work	Time Frame ≤ 6 months
5	Reducing costs in operations.	• Reduce inventory costs via improving the JIT/Kanban program & continuous flow. • Review and optimize line layouts and productivity. • Introduce Cell manufacturing for low-volume lines. • Reduce total line defects <0.3%. • Work with suppliers to reduce incoming defects and costs	Time Frame ≤ 6 months
6	Reduce product failure rate.	• Conduct repeat PDCA training. • Kaizen activity: Select high fail products, and improve manufacturing process to reduce internal and external fail rate	Time Frame ≤ 6 months
7	Accelerating JIT/Kanban to reduce WIP and overall inventory.	• Review and increase kanban activity with all suppliers. • Introduce JIT and Kanban system on all production lines • Continuous flow on final assembly lines	Time Frame ≤ 6 months

The Business Plan Approach: Using Hoshin Planning

An alternative approach to getting started is to integrate the transformation into the annual plan. If your entity has an annual business plan, then it is appropriate to consolidate the methodologies into the plan. It could be your operation's specific planning process or the Hoshin planning process.

We have discussed the Hoshin planning cycle and stated that a good planning process will ensure you do the right things, such as: review customer needs, evaluate company needs, set priorities, and define plans for improvements and increasing revenue. In the General Manger's plan in Figs. 10-5 and 10.6, the objective was to increase revenue and profits. This required aggressive strategies, are tabulated in Table 11-1 for your reference. Some comments on the plan in Table 11-1 are appropriate:

- The focus is on resolving a critical business issue. There are two objectives: Increase revenue and profits.
- The manager has selected the appropriate tools which will yield the best results. Some of the strategies will be from our tool box. In this case, JIT/Kanban and continuous flow can be rolled out throughout the operations. This will also require a full review of cycle time and standard work at all production lines.
    o Within this strategy the team can use several tools and methodologies: Get a team to take the gemba walk approach and identify opportunities; then do the analysis, use the appropriate methodology, and rollout improvements. Alternatively, use VSM mapping, do the analysis, use the appropriate methodology, target the future state, and rollout improvements

This approach allows you to integrate the relevant methodologies into the organization's business plan of increasing profits; now you are completely synchronized with the organization's objective and direction.

- If the management team is well-educated in the tools, plus understands both customer and internal issues they could come up with more aggressive strategies than those we have suggested in Table 11-1.
- The entire operation works together towards a common goal of improving operations with the end goal of increasing revenue and profit.
- With a detailed business plan, you will be reviewing progress and results during regular reviews. During the reviews, you will be able to adjust and tune your strategies for best results.
- On a yearly basis, you can introduce or improve selected methodologies. But you must do them well, you must plan and measure progress and the entire team must be focused to make things happen.

- With experience, you can get more aggressive in transforming the business and increase the pace of change.

Our preference is clearly for the Business Plan Approach, with its focus and deep integration into the company's business, direction, processes, and customers.

Challenges and Solutions for Success

We have gone through the tools and methodologies that help achieve operational excellence and we have discussed how to get started. Nevertheless, there are many challenges in transforming an operation towards the new direction.

In fact many operations that start a lean and TPS program do not get the expected results. A survey conducted by *Industry Week*[48] found that only two percent of companies that have a lean program achieved their expected results. Furthermore, less than a quarter of all companies surveyed reported significant results.

Recently the Shingo Prize Committee[49], which gives awards for operational excellence, checked with past winners and found that many had not sustained their progress after winning the award.

Why? We discuss the most critical challenges and provide some solutions from our personal experience. This will also serve as a guide to the CEO or senior manager who wants to implement change and take the journey towards operational excellence.

Long term management commitment: This is by far the greatest challenge. Results cannot be achieved over the short term. Hence, the management team, per Dr. Deming, must have *"constancy of purpose "* over the long term in driving towards the company vision and direction. The initial commitment requires a leap of faith.

Be patient; if you are starting from a low base, then major improvements and impact may take a few years. Pockets of improvements will be visible but major improvements across the organization will take time. Therefore it is crucial to measure progress and to document the benefits and savings and convey these results to the management team.

Toyota started TPS in the machine shops and that is where the ideas of Just-In-Time manufacturing and quality via the Jidoka system came from. It was a slow and demanding process but once learnt it spread throughout the company. It will be hard work, but once you see initial results you will be motivated to move faster. Remember Toyota took a long time to achieve today's state of excellence. The Vice-Chairman of Toyota said this of the Toyota Way: *"I don't think I have a complete understanding even today, and I have worked for the company for 43 years."* However, with lessons learned from Toyota, all the documented best practices and tools, and an experienced Sensei to guide you, you will make good progress.

If you decide to introduce the Hoshin planning process to drive your efforts, expect at least two years to have an be acceptable Hoshin process; and at least three years to be proficient in the methodology. The same applies if you plan to use TOC methodology.

Be aware of insufficient preparation and learning. After one year of struggle, you may decide that "we tried it but it doesn't work here" and give up. Dr. Noriaki Kano has called this the *false learning curve* – it may take two or three years of effort before you can see positive improvement across the organization. In the mean time, you must be patient, humble, and persistent; such temporary setbacks will soon pass.

Be prepared to ensure and implement cultural change: As the tools and projects are implemented, the team must accept and implement cultural change. Change here means continuous improvement in everything you do, understanding and meeting customer needs, continuously upgrading operator and engineer skills, and allowing decision making to move down the operation to project teams, engineers, and junior managers. This is done best in the context of the business plan approach, and has to be repeated and improved yearly.

The basic infrastructure must be in place. For any good manufacturing operation, the basics must be in place: This includes a strong team – lean manufacturing does not imply you can succeed with fewer resources. It is a fallacy to expect miracles with skimpy resources. This includes a good network of suppliers delivering decent quality of parts, strong purchasing department, a trained workforce, and competent engineers to resolve issues and bottlenecks as weak processes are exposed and improved.

Focus your efforts. We do not propose that you *"Implement TPS or Lean Manufacturing throughout the operation"*; such an approach is too grand and difficult to implement given you are also running a business and will have many competing priorities. New methodologies must be selected based on customer and internal company needs. Hence start with, say, higher quality and fewer defects – because these are the company attributes which will be immediately observed by customers and help sales. If your quality is very good, go down the list of methodologies and select as appropriate based on business needs. For example: Higher productivity, quicker response time, reduced inventory, and lower costs. With time you will be able to introduce all the tools discussed. Gains in these processes must be sustained, via daily management plans or KPIs.

Understand, learn, and practice the tools. Remember that the tools are a set of best practices that have been improved and documented, in this text and elsewhere. But you cannot apply them blindly via a blitz and expect immediate results. Hence, look at your current practices, understand why some are not working to your expectations, eliminate the root causes and adopt and adapt the new tools to fit your situation; then provide training and ensure the team has a deep understanding of the underlying principles of the tools. Only then can they implement the tools successfully; apply them equally well elsewhere, and further improve the system.

Strike a balance between costs and value. TPS and lean manufacturing started on the shop floor with the goal of reducing waste and cost. But these tools have evolved into providing value to the customer; value such as:

- Quicker delivery, via faster response time.
- Deliver only what is required, just in time, and in smaller batches.

- Excellent quality – which includes products that seem to last forever.

Operational excellence will lower your cost structure and allow your company to add this value without increasing prices, thus making the final product more attractive and competitive. To benefit from operational excellence, it is important to add value as costs go down. This is the cost-value equilibrium[50] as defined by Hines and Holweg of the Cardiff Business School; this concept is illustrated in Fig. 12-1.

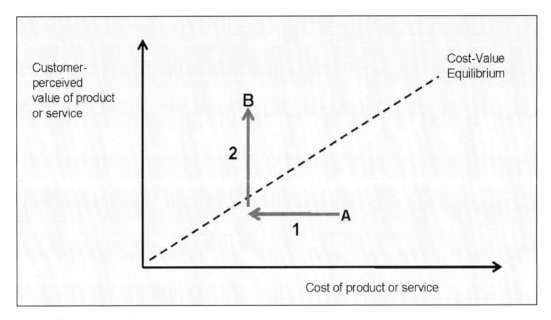

Fig. 12-1: The Cost-Value Equilibrium.
The figure illustrates the concept of reducing waste and cost (Step 1) and adding value (Step 2). This allows a company to move from point A to B; but sometimes a company may prefer to emphasize only on cost reduction, that is: do step 1 only.

Summary

We have come a long way. We have discussed the tools and best practices of TPS and Lean Manufacturing. Some of the tools will get you organized. Tools such as Standard Work, Managing Cycle Time, JIT, and Continuous Flow manufacturing will improve productivity and quality. We discussed several quality tools that will help you aim for zero defects. Value stream mapping will help to identify and remove waste, hence creating value for customers, while the PDCA approach will provide the structure to mange improvement projects.

We also discussed employee development and how to aim for zero defects and Total Quality Creation. All these activities can be driven via Hoshin planning, or an alternative such as TOC or good project management. Finally, we discussed how to accomplish the transformation towards Operational Excellence.

Managers and employees must learn these principles and best practices – and absorb them into their daily behavior. The road ahead is tough and challenging. But focus and persistence will give excellent results. A vital point to remember is that the goal is one of continuous improvement, with new systems and methods replacing old ones.

Operational excellence is a long journey but it is not the destination or goal.

Once started, you will be on an unending journey of waste reduction, continuous improvement, and manufacturing excellence by all resources in the company with the goal of increasing wealth in your corporation and providing the highest quality and value to your customers.

Appendices

Appendix 1: Glossary of Terms

Andon: Japanese for lantern or light. It refers to a flashing light on production equipment or line that alerts the operator or technician that a defect has been discovered and needs attention.

Autonomation: This word was coined and made popular by Shigeo Shingo, of Toyota; he believed that the separation of man and machine was important to ensure high quality. The term is equivalent to jidoka, and refers to intelligent machines that can detect abnormalities and stop the machine plus alert the operator.

Continuous flow: A manufacturing policy of producing one-piece of product at a time in a continuous flow process.

Countermeasures: Corrective action taken to solve a problem. The two expressions are used inter-changeably.

Cycle time: This is the time taken for 1 part on 1 machine to complete 1 cycle from start to finish (excluding changeover time, wait time, or queue time).

Changeover time: When production switches from running one product or part to another product or part, there will be *changeover time* to configure and prepare for the change. The changeover time is measured from the last part of the previous batch, to completion of the first good part in the current batch. The changeover time will encompass any machine setup time in production.

Five S or 5S: A system for workplace organization. The 5S system refers to 5 activities beginning with the letter 'S", namely Seri (or Sort), Seiton (or Stabilize), Seiso (or Shine), Seiketsu (or Standardize), and Shitsuke (Sustain).

Gemba: Japanese for "the place", or where it is happening. The gemba approach – often called "gemba walk" – is about making decisions after observing work on a production line and getting your hands dirty.

Genchi genbutsu: Japanese for "go and see for yourself". This is similar to the expression "gemba walk" in order to observe work on a production line.

Heijunka: Heijunka means "to smooth" in Japanese, and refers to the product leveling process that helps to manage a constant production rate in the upstream assembly area. Typically this is done manually via a board or a box, with many pigeon holes; the X-Axis represents time and the Y-Axis representing the product in queue;

each box could represent one hour time slot. The schedule or kanban cards are placed in the holes as per the downstream or customer order.

Hoshin Kanri: Hoshin means objectives or directions, while Kanri means control or management. So in essence Hoshin Kanri planning means *Policy Management* or *Management of Objectives*.

Jidoka: The principle of Jidoka describes machine capability which has intelligent design to stop a manufacturing process whenever abnormalities occur. This line stop can also be done manually by a production operator whenever a defect is detected. This was popularized by Shigeo Shingo, of Toyota, who coined the term *autonomation* to mean intelligent machines that can detect abnormalities and stop the machine.

JIT or Just-in-Time: Refers to the just-in-time production process which manufacture only what is needed, when it is needed, and in the amount needed.

Kaizen: Kaizen is Japanese for improvement. Hence, the term kaizen is used to represent a philosophy of continuous improvement in manufacturing or business.

Kanban: Kanban means signal or card. Kanban is a scheduling system that coordinates production and withdrawals to ensure just-in-time production. The original kanban system used cards as signals to move inventory; nowadays there are many alternatives that perform the same function.

Mizusumashi: This is Japanese for water-beetle. A water beetle moves back and forth erratically on the water surface; this behavior personifies what the MH is doing all day- running back and forth on the factory floor.

Muda: Japanese for waste or an activity that is wasteful. Refer to the discussion on seven wastes in the text.

Nichijo Kanri: Nichijo means Daily. Hence Nichijo Kanri means *daily management* of routine processes in manufacturing or a business.

PDCA: Plan – Do – Check – Act. This refers to the scientific or structured process used to manage and improve an activity.

Poka yoke: Japanese for setting up a mistake-proof or fail-safe procedure in manufacturing.

Quality definition: *Products and services that meet or exceed customers' expectations.* The key words here are exceeding customer expectations, which can be very demanding and will be relative to what other manufacturers or companies provide.

Sensei: Japanese for teacher or master. In manufacturing, a sensei is a very experienced manager or engineer who understands TPS and lean manufacturing. He or she is able to train others in the techniques of finding waste and problems, and subsequently to use the correct tools to resolve the issues.

Setup time: This is the time taken to replace the existing die, jig, or fixture with a new one, for production of the next part. This is a subset of changeover time.

Standard work: Standard work documents the current work instructions or methodology for building a product in the most efficient way with the best quality. It will include all the steps in the work sequence, the cycle time required to complete the work, and the inventory required at the workstation.

Takt time: Takt is a German word for the beat of music. This was adopted by German manufacturing as *taktzeit*, which translates as cycle time. Takt time aims to match the pace of production with customer's demand; hence in manufacturing takt time sets the pace for production lines. Therefore, the time needed to complete work at each workstation has to meet or be less than the takt time goal.

TOC: The theory of constraints (TOC) was expounded by Dr. Eli Goldratt. He puts forward the idea of a system constraint that hinders progress or profit in a business. His theory comes from German manufacturing theory of managing takt time, which strives to continuously remove the weakest link in a series of manufacturing steps, in order to remove bottlenecks and speed production. According to Goldratt, the same concept applies to business; hence we must identify and fix each constraint in order to improve profits and business performance.

TPS: Toyota Production System. TPS encompasses the philosophy and practices that Toyota Motor Company has developed over the years. The TPS lays the foundation for lean manufacturing.

TPM: Total Productive Maintenance. The objective of TPM is to ensure that machines provide predictable performance, deliver exactly what we expect of them, with zero errors, breakdowns, or accidents; all this at a reasonable cost.

TQM: Total Quality Management. The goal of TQM is continuous quality improvement and high customer satisfaction. It promotes management and worker participation to achieve this goal via tools like the PDCA cycle, quality circles or kaizen teams, and Hoshin planning.

Value Stream Mapping: Value stream mapping (VSM) is a tool to highlight and remove waste in an organization. The VSM map charts the current activities and processes in an organization and identifies both value and waste. The opportunities for improvement can be identified from the VSM map and the future state and improvement plans can be developed.

Work Instruction: These is a sub-set of standard work, and shows the specific work sequence at a workstation and the assembly material required. It's often displayed at each workstation instead of the more detailed standard work.

Yamazumi Chart: Yamazumi is Japanese for heap, mound, or stack; hence a Yamazumi chart is a stacked bar chart, which gives a visual display of individual tasks and cycle times at various stations on an assembly line or cell Vs the takt time.

Appendix 2: Assessing Operational Excellence

To help identify opportunities for improvement, we have provided an assessment matrix which will allow you to review strengths and weaknesses of your factory operations and identify opportunities for improvement.

In the next few pages is an assessment matrix for your review. To get started, you should appoint a team from production, engineering, and management. Review each item and try to determine where you are today; in most cases you will need to go and get data to understand where you are today. Hence, go over the items, assign responsibilities and agree to meet again. It will take several meetings to get a good measurement. This is only a rough guide but it will get you started to think and question your activities and priorities.

Although the matrix allows for an overall assessment score, that is not our intention – the purpose is to identify opportunities and priorities. This matrix can be downloaded from the opex website: www.opex-usa.org

Table A-1: Assessment of Operational Excellence and Factory Performance		Assessment and Scoring (Page 1 of 3)			
	Item Reviewed	1: Unacceptable	2: Minimum requirement	3: Good	4: Very Good
1	**Customer metrics and performance**				
a	Customer complaints, CARs, and issues are registered and resolved quickly.	Average time (AT) >10 days	AT <10 days	AT <5 days	AT <2 days
b	How is delivery to customers? Are you meeting commitments? Other customer metrics should be in the KPIs, next item.	<95% Ontime	>95% Ontime	>98% Ontime	100% Ontime
2	**Factory metrics and performance**				
a	Are there robust metrics to measure factory performance? Key metrics must include: *Product quality and product yields *Delivery to customers *Cost management & cost reduction	None	Some KPI or metrics are available.	Good set of metrics covering quality, cost, delivery, yields.	Good set of metrics covering quality, cost, delivery, yields.
b	The KPIs reflect a direction and philosopphy of continuous improvement - based on past and future targets. This requires a past and current target review.	No KPI or metrics.	KPIs exist, but no year to year improvements	Several KPIs have improved.	Improvements achived in > 5 metrics
c	There is good performance to above metrics	None	Missing many targets	Missing a few targets	Meeting all targets
d	The operation is working on continuous improvement and reduction of process variability: For example incoming supplier defects, batch production, weak TPM, long machine setup times, high WIP, incoming vendor defects, unbalanced production lines, excessive overtime.	No effort.	Some reviews, with marginal gains.	Regular reviews, with good progress in a few areas.	Regular reviews, with good progress in listed areas.
3	**The organized workplace - 5S system**				
a	Is the workplace clean & spotless, with no clutter, dirt or debris?	Unacceptable	Some effort, decent gains.	Good effort, can do better	Tip-top condition
b	Is the workplace organized: Workstations, equipment, and tools are acessible to operators, with no clutter and unnessecary items lying around.	Unacceptable workarea	Disorganized, needs much work.	Good layout, but cluttered, needs improvement.	Well-organized workplace.
c	Workplace is well laid out, workflow is streamlined and efficient; areas are marked with good signage; standard work instructions are visable and properly located; the layout supports kanban.	Disorganized workplace. Needs major revamp.	Workplace layout is mediocre, needs improvement.	Good layout; some congestion, minor issues.	Very good layout; no issues at all.
4	**Standard work**				
a	Standard work instructions at all workstations. Either paper documentation or via monitors.	None visable	Poor displays or missing in assy. lines.	Complete in older lines but missing in new assy. lines.	Complete.

Table A-2: Assessment of Operational Excellence and Factory Performance

Assessment and Scoring (Page 2 of 3)

	Item Reviewed	1: Unacceptable	2: Minimum requirement	3: Good	4: Very Good
4	**Standard work (Continued)**				
b	Standard work instructions are complete and accurate per actual work practice. Urgent changes require temporary documentation or red-lining by engineers to ensure operators are updated on standard work. Standards are reviewed for accuracy via a QA audit.	No checks or audits	Many errors or obsolete standards.	Mostly updated and accurate. Some red-lined documents, not updated.	Extensive audits: 100% accurate.
c	Work processes are continuously improved as evident in improved and re-documented standard work; check per engineering records.	No recorded effort.	Some Improvements	A reasonable effort to improve.	A planned effort with continuous improvement.
5	**Identification and reduction of inventory waste**				
a	There is a minimum work-in-process inventory in production: Measured by hours of WIP.	>48 hrs	<48 hrs	<24 hrs	<12 hrs
b	There is a minimum of inventory in the warehouse: Measured in turns; 12 turns/year = 12/12 = one month or 30 days of inventory. Other waste is identified below in workflow and quality sections.	<6 turns	6 to 12 turns	12 to 30 turns	>30 turns
6	**Work Flow**				
a	How is the factory layout determined? Is the current layout equipment based or product based?	All equipment based	Mostly equipment based + some product based	Mostly product based + some equipment based	Fully product based.
b	Takt & Cycle time is managed well and lines are balanced and flowing smoothly. There are no choke points or bottlenecks in work cells or assembly lines.	Operation is run in batch production mode.	Bottlenecks are common in cells or assembly lines.	Some bottlenecks, mostly smoooth flow	Smooth flow in both cells and assy. lines.
c	Production is run in JIT (Just-in-time) mode, with a Kanban material delivery system. Note: Refer to results of items 5 (WIP) and 3 (5S system) to validate effectiveness of JIT.	None	JIT and Kanban on >50% of production.	JIT and Kanban on >80% of production.	Fully JIT amd Kanban
d	Is there a continuous flow on the production line? For both cell manufacturing and linear paced lines, expect smooth flows with little batch production..	Completely batch	Batch plus flow	Some batch production, mostly flow	Very smooth flow
7	**Machine Management & TPM**				
a	There is a strong equipment maintenance program; hence no equipment breakdown impacting production flow or causing downtime.	No TPM effort	Basic TPM program, weak implementation, frequent breakdowns.	TPM program exists but often misses schedule. Little downtime.	Strong, effective program. No production downtime.
b	Equipment availability is good: Uptime or OEE is measured and targets met. Process capability, Cpk targets, achieved for all eqiuipment.	Not measured	Uptime over 90%; No OEE; Cpk targets for some equipment only.	High OEE 80-90% range; meet Cpk targets for most equipment.	OEE >90%; Meet & exceed Cpk meets targets for most equip.

Assessing Operational Excellence   259

Table A-3: Assessment of Operational Excellence and Factory Performance		Assessment and Scoring (Page 3 of 3)			
	Item Reviewed	1: Unacceptable	2: Minimum requirement	3: Good	4: Very Good
7	**Machine Management & TPM (Continued)**				
c	Does equipment produce defective parts due to poor maintenance?	Not measured	Often	Sometimes	Never
d	**SMED:** There is a strong effort to reduce equipment setup times for pressess, molds, SMT, testers, line setups, etc. Check the data and the effort to reduce changeover times for production lines and equipment over the years.	No effort.	SMED in <30 min. for some equip. and line setup.	SMED in 10-15 min. range, for most equip and line setup.	SMED in 1 to 5 min. range for most equipment, and <10 min for balance.
8	**Quality:**				
a	Good supplier quality – dock to stock inventory due to supplier certification and management.	None	A few incoming parts, >50%.	Most incoming parts, >90%.	All incoming parts
b	What are the first-pass product assembly yields? Is the production defect rate high? Are there pallets and shelves of product awaiting rework?	<90%	90-95% range for most products.	>95% for most products.	>99% for some, and >95% for balance.
c	There are extensive poka-yoke and jidoka systems in place on the factory floor for critical processes.	None	Done on a few lines	Done on >75% of lines	Implemented on all lines
d	Operation runs smoothy with no fire-fighting. Check for 3 modes: Fire-Fighting mode; Systematic Problem Solving Mode; Predict & Prevent mode. *This measure is an indicator of effective managemnt, strong engineers, and good problem solving skills.*	Operation is in fire-fighting mode, yields are low	Mostly fire fighting. PDCA problem solving used in some areas	Strong use of PDCA: Systematic Problem Solving Mode	Operation is beyond PDCA: Predict & prevent mode
f	Theree is a good improvement process in the operation: Improvements are mesuable in in processess, cycle times, quality, product warranty.	None	Some in several areas	Evident in most areas	Extensive in all areas and measurable.
9	**Employee/Partner Participation and Development**				
a	Establised and effective skills-training program for operators. Routine training for all operators, well documented; training effectiveness is measured.	None	Poor documentation and sporadic implementation.	Well documented, but often missing training schedules.	Well documented, with good implemention.
b	Schedule for Training for professional staff: Including PDCA, 7 tools, DOE/Six Sigma, poka yoke, and other training.	None	Poor documentation and sporadic implementation.	Well documented, but often missing schedules.	Well documented, with good implemention.
c	Partners and suppliers: 1) Have been trained in company needs and requirements; 2) Help to improve product quality and delivery; 3) Particpate in kanban delivery.	None	Success in 1 area	Success in 2 areas	Success in all 3 areas
10	**Strategic direction of operation/factory**				
a	Senior management has a strategic plan for the next few years to ensure competitiveness of the company. The plan covers: operation growth, productivity improvements, cost structure improvement, people management. There is visible and measureable progress in this plan.	No plan	These are some ideas which are executed randomly.	Some planning with execution.	Strong, detailed, thoughtful plan, well executed.

Appendix 3: References

We provide here a list of books for further reading.

5 Pillars of the Visual Workplace: By Hiroyuki, Hirano, Productivity Press, New York, 1990. This is a very detailed and complete text about the 5S system.

A study of the Toyota Production System from an Industrial Engineering Viewpoint: By Shingo, Shigeo, Japan Management Association, Tokyo, 1985. This is a classic written by an engineer who helped develop the Toyota Production System. However it is not an easy read – you have to search for the gems.

Creating Level Pull: A Lean Production-System Improvement Guide for Production-Control, Operations, and Engineering Professionals: By Art Smalley, Lean Enterprise Institute. This text provides a practical guide on material and kanban management in a leveled production environment.

Guide to Quality Control: By Kaoru Ishikawa, Asia Productivity Organization. This is also available in the US from the ASQC Quality Press. It's a paperback with a good detailed review of 7 tools and SPC/SQC, and will be a good reference for the engineers.

Lean Thinking: Banish Waste and Create Wealth in Your Corporation: By Womack, James; Jones, Daniel; 1996, Simon & Schuster, New York. The authors provide a thoughtful discussion of their value-based business system based on the Toyota model. The ideas in the text will get you thinking big.

Learning to See: Value Stream Mapping to Add Value and Eliminate Muda; By Rother, Mike and Shook, John. Per the title, this is a guide for improving productivity via VSM; the VSM tool helps to identify and eliminate non-value activities.

Logical Thinking: By Dettmer, W.H., Quality Press; this text gives a very detailed discussion of TOC and how it can be applied to effective planning for a company. However, Dettmer tends to get heavy on theory; hence you will find Dr. Legat's less complicated approach of merging TOC with Hoshin planning very useful. Refer to Chapter 10.

Out of the Crisis: Edwards Deming, MIT Press. A classic, by the redoubtable Dr. Deming; here he explains his philosophy, via his 14 points. Not an easy read, but nevertheless a must for all managers and professionals.

Quick Changeover for Operators: The SMED System, Productivity Press; A good concise book on the how and why of quick changeovers for machines on the shop floor. It's written to be used as a training manual.

Statistical Quality Control Handbook*:* Published by Western Electric. This is an excellent easy to read book, on Statistical Process/Quality Control. The discussion focuses on control charts, design of experiments and process capability. This book may be out of print.

Statistical Quality Control*:* E.L. Grant and R.S. Leavenworth, McGraw-Hill, New York. Engineers who want to understand the theory and practice of SPC/SQC will like this book.

The Goal: A Process of Ongoing Improvement: Goldratt, Eliyahu; 2004, North River Press: "The Goal" is an entertaining novel and a thought provoking business book. The book provides a good introduction to the Theory of Constrains (TOC).

The Toyota Way: By Liker, Jeffrey, 2004: McGraw-Hill, New York Author Jeffrey K. Liker's thorough insight into the continual improvement method known as "The Toyota Way" reflects his experience with the Toyota Production System (TPS) and his knowledge of its guiding philosophies and its technical applications.

Today and Tomorrow*:* Ford, Henry, 1988; Productivity Press, Portland, Oregon. Many of Henry Ford's ideas and practices found their way into the Toyota Production System and lean manufacturing methodology.

Appendix 4: Resources and Websites

Most templates used in this text are available for download at: www.opex-usa.org
- Hoshin planning forms
- PDCA cycle and A3/A4 process templates
- Suggestion scheme forms
- Standard work forms and templates
- Lean/TPS Assessment matrix
- Completed projects
- Alternative websites to get tables and formats for Hoshin planning, standard work, Yamazumi charts, etc. are also listed.
- Instructive videos can be accessed via the links at our website. These include Jidoka, Andon, Chaku-chaku, and other videos.

Websites that provide information on topics discussed in the text:

Deming's 14 points can be viewed at several websites, including: http://en.wikipedia.org/wiki/W._Edwards_Deming http://www.youtube.com/watch?v=ehMAwIHGN0Y&feature=related

LEI (Lean Enterprise Institute): LEI has an excellent website with various resources, books, seminars, and forums. Founded by the formidable James Womack, its mission is to "Advance lean thinking throughout the world." Website: http://www.lean.org

Society of Manufacturing Engineers has books and videos on many lean manufacturing topics; membership required. Website: www.sme.org

TOC (Theory of Constraints): More information at the Delta Institute on using TOC for planning: http://www.delta-institute.com

TWI (Training within Industry): Several websites offer programs on skills-training methods. For more information check: http://trainingwithinindustry.net/index.html and http://www.twi-institute.com/

Hoshin planning tables and scorecard from PlanBase Inc. This excellent website provides Hoshin and scorecard templates to automate the planning process via an intuitive, click and go application. It can rapidly deploy plans in a cascading linked fashion across an organization, ensure that strategic plans are communicated and measured across the organization and reduce the time required for individuals and teams to update performance. Website: http://www.planbase.com

Index

A

A3/A4 process · 103, 127
   case study · 128
Akers, John · 125
Andon · 149, 253
Assessment of Operational Excellence
   assessment checklist · 241
   conducting an assessment · 240, 241
Autonomation · 147, 253

B

Bennis, Warren · 193
Bottlenecks · 59
Boyd, John · 124
Bridgestone · 194
Build to order · 62
Business fundamentals plan · 195
Business Plan approach · 231, 239, 247

C

Carnegie, Andrew · 179
Cause and effect diagram · 112, 118, 124, 226
Cell production
   benefits · 168
   types of · 172
Cell system layouts · 170
Changeover time · 159, 253
Company or organization vision · 194
Competitive benchmarking · 106
Containment action · 113
Continuous flow · 155, 253
   converting from batch to continuous · 174
Continuous process improvement · 180
Control charts · 136
Corrective action
   types of corrective action · 113
Cost-value equilibrium · 250
Countermeasures · 253

Cpk of machines · *See* TPM
Csikszentmihalyi, Mihaly · 155
Culinary Institute of America · 14
Customer inputs · 195
Customers first · 151
Cycle time · 29, 45, 51, 253
   definition · 33, 48
   effective · 59
   total · 46
   unbalanced · 159
   Vs overall manufacuring cycle time · 59

D

Daily Management plan · 194, 195
   elements and guidelines · 215
Decision rules
   for cell and linear production · 164
Defect detection · 141
Defects and poor quality · 6
Deming prize · 194
Deming, Edwards · 103, 180
Design of cellular system · 168
Design of linear system · 165
Dettmer, William · 224
DFM or design for manufacturing · 132
DMAIC methodology · 124

E

Economic order quantity or EOQ · 66, 83
Education and training · 189
Eight D or 8D process · 123
Einstein, Albert · 27
Ellis, Henry · 83
Emerson Electric · 193
Employee education · 179, 181
Employee suggestion scheme · 179, 183
   objectives and guidelines · 184
   reward system · 185
   success factors · 187

Excess inventory · 6

F

Facilitator, Team · 180
Fire fighting · 132
Five S or 5S System · 9, 11, 253
    5S audit checklist · 21
    5S audit process · 20
    5S barriers · 23
    5S committee · 18
    5S display board · 17
    customer expectations · 26
    good 5S practices · 17
    measure 5S progress · 22
    objectives of 5S · 10
    Seiketsu · 11
    Seiri · 11
    Seiso · 11
    Seiton · 11
    Shitsuke · 11
Five W, 5W, or 5W/2H method · 111
Ford, Henry · 4, 28, 45
    CANDO process · 9

G

Gemba · 37, 253
Genchi genbutsu · 37, 238, 253
Goldratt, Eli · 225, 227
    The Goal · 124, 225

H

Heijunka · 253
Heijunka process · 77
Hewlett-Packard · 116
Hirano, Hiroyuki · 9
Hokake, Takoshi · 153
Hoshin Kanri planning · 63, 179, 193, 194, 254
    criticism of · 224
    illustration · 197
Hoshin plan
    deploying objectives · 208
    deployment matrices · 209, 210
    flowchart · 194
    formats and guidelines · 199
    guidelines for a review · 219
    implementation plan · 213
    performance measures · 207
    QCDE · 200
    reviews · 197, 218
    strategy · 205
    super hoshin plan · 227
    Vs Daily Management plan · 199
Human errors · 144

I

IBM · 183
Improvement project · *See* PDCA cycle
Improvement projects
    types of · 125
Improvements projects · 181
In Process Quality Audits or IPQA · 137
Ishikawa diagram · *See* Cause and effect diagram
Itoh time management model · 223

J

Jidoka · 147, 148, 254
JIT · 65
    challenges · 82
Just-in-time production · *See* JIT and kanban

K

Kaizen · 254
Kaizen and kaizen teams · 125, 179
Kaizen teams
    management · 180
    recognize and reward · 181
Kanban · 65, 67, 254
    aim of kanban · 82
    challenges · 81
    description of kanban · 66
    getting started · 74
    kanban improvement cycle · 70
    kanban system rules · 69
    supplier kanban · 80
    two-card kanban, one-card, and cardless · 71
    with suppliers · 190
Kano model · 105
Kano, Noriaki · 104, 134
Knight, Charles · 193
KPI or key process indicators · 195
Kume, Hitoshi · 144

L

Law of Variability · 177
Law, Little's · 155, 160
   and continuous flow · 161
Leadership · 229
Lean Manufacturing · 3, 7, 189, 193, 238, 240, 242, 249, 250
Leveled production · 61
Levitt, Ted · 106
Levitt, Theodore · 106
Liker, Jeffrey · 2, 179
   4P model · 2, 179
Line balancing project · 52
Long-range plan · 194

M

Matsushita Electric · 183
McDonalds · 14
Metrics for measuring machine failure · 93
Miltenburg, John · 162
Mistake-proofing · *See Poka Yoke*
Mizusumashi · 69, 72, 254
Model change notification · 42
Muda · 3, 5, 33, 254

N

Nemawashi · 127
Nichijo Kanri · 194, 254
Numerical targets · 217

O

OEE · 98
Ohno circle · 40
Ohno, Taiichi · 9, 27, 40, 65, 82
One-piece production · 155
   benefits · 158
   challenges · 158
OODA loop or observe-orient-decide-act · 124
Operational excellence · 155
   an unending journey · 251
   best practices · 3, 6
   challenges · 248
   definition of · 7
   why do it? · 240
Operator certification · 44

Overall equipment effectiveness or OEE · *See* OEE, *See* OEE
Over-processing · 6
Overproduction · 6

P

Paced linear line · 165
Pacing production lines · 49
Packard, David · 226
Partnering with suppliers · 179, 190
PDCA cycle · 103, 107, 123, 132, 153, 227
   A3/A4 process · 127
   completed project · 116
   detailed · 109
   figure of · 107
   improvement or control? · 133
   management · 103
Performance measures · 207
Planning cycle · 222
Poka yoke · 145, 254
   categories of · 145
Policy deployment · 193
Prevention of problems by prediction · 132
Preventive action · 113
Problem solving hierarchy · 132
Problem solving styles · 107
Process and quality control · 134
Process capability · 96
Process-FMEA · 132
Production system · 161
   cellular · 163
   direction of workflow · 173
   equipment-based · 162
   linear · 163
   product-based · 163
Project Management approach · 231, 239, 243

Q

Quality · 103
   definition · 104
   product and service · 103
Quality fatigue · 43, 49
Quality first · 152
Quality goals
   setting · 126, 150
Quick setup of machines · 83

R

Red-tag process · 12
Reducing production lead times · 83

S

Sensei · 238, 242, 254
Setup time · 83, 254
    improvement project · 87
    reduction · 85
Setup time Vs changeover time · 87
Seven tools · 124
Seven wastes · 5, 6
Shingo Prize for Operational Excellence · 2
Shingo, Shigeo · 69, 145, 147
Siemens · 183
Single Minute Exchange of Die · *See* SMED
Six-Sigma · 2
SMED · 84, 87, 89, 101
Spaghetti chart · 36
Spaghetti lines · 36
Standard work · 27, 254
    contents of · 28
    good and bad · 38
    improving · 35
    preparing · 31
    sample of · 29
Statistical process control · 135
Strategy · 205
Successive inspection system · 147
Systematic problem solving · 132

T

Takt time · 45, 255
    definition · 46
    Vs cycle time · 47
Taktzeit · 45
Taylor, Frederick · 28
Ten X or 10X rule · 141
Theory of constraints · *See* TOC
TOC · 225, 255
    improvement cycle · 124
    Super Hoshin plan · 227
    Vs Hoshin planning · 228
Total participation · 152
Total Product or Service concept · 106
Total Productive Maintenance · *See* TPM
Total Quality Management · *See* TQM

Toyota · 2, 28, 65, 127, 137, 147, 183, 238, 239
Toyota Production System · *See* TPS
TPM · 2, 90, 101, 255
    goal · 91, 94
    Machine failure pattern · 90
    plan · 90
    process capability · 96
    requirements for · 93
TPS · 2, 3, 5, 7, 62, 165, 238, 240, 248, 255
TQC or Total Quality Creation · 132, 152
TQM · 103, 151, 255
    principles of · 151
Training process · 43
Training within Industry · 43

U

Unnecessary motion · 6
Unnecessary transportation · 6

V

Value stream mapping · *See* VSM, *See* VSM
Variability in manufacturing · 174
    Law of variability · 177
    minimizing · 174
Variability of processes · 61
Visual factory, The · 13
VSM · 244, 255
    map · 233
    procedure · 232
    projects · 235, 237

W

Waiting time · 6
Whispering of Satan · 153
Womack and Jones · 3, 5
Work combination sheet · 33
Work instruction · 255
World-Class manufacturing · 132, 155, 177, 191

Y

Yamazumi chart · 57, 255

Z

Zero defects · 103, 140, 141

Notes and References

Chapter 1

[1] These attributes are typical to the Toyota Production System. Within some attributes we have used points gleaned from the outstanding piece by R. Eugene Goodson, *Read a plant – Fast*, Harvard Business Review; May 2002. Mr. Goodson comments that the trained eye can read a plant in 30 minutes and tell if it is truly lean. This is a good read; you can access it if you Google the article title.

[2] The *Shingo Prize for Operational Excellence* is named after Shigeo Shingo, who worked at Toyota Motors and helped develop the Toyota Production System. The model and award is managed and hosted by the Jon M. Huntsman School of Business, Utah State University, USA.

[3] Liker, Jeffrey, 2004: *The Toyota Way,* McGraw-Hill, New York.

[4] Womack, James; Jones, Daniel; 1996, Lean Thinking: Banish Waste and Create Wealth in Your Corporation, Simon & Schuster, New York.

[5] Ford, Henry, 1988: *Today and Tomorrow,* Productivity Press, Portland, Oregon.

Chapter 2

[6] Taylor, Frederick W., 1911, *Scientific Management:* Available free online at http://www.gutenberg.org

[7] Ford, Henry, 1988, *Today and Tomorrow:* Productivity Press, Portland, Oregon.

[8] Ford, Henry, 1922, *My Life and Work:* Available free online free at http://www.gutenberg.org

[9] Hirano, Hiroyuki, 1990, *Pillars of the Visual Workplace:* Productivity Press, Portland, Oregon. This is probably one of the most detailed texts on understanding and implementing the 5S system.

[10] I am grateful to Jeremiah Josey for providing these pictures and other details of 5S practices.

Chapter 3

[11] Taylor, Frederick W., 1911, *Scientific Management:* Available free online at http://www.gutenberg.org

[12] Ford, Henry, 1988, *Today and Tomorrow:* Productivity Press, Portland, Oregon.

[13] Translation from Wikipedia: *The Online Encyclopedia.*

[14] Adapted from: Soin, S., 1998, *Total Quality Essentials:* McGraw-Hill, New York.

[15] Quoted in: Liker, Jeffrey K, 2004, *the Toyota Way*: McGraw-Hill, New York.

[16] Source: War Production Board, Bureau of Training Within Industry Service, 1944, Job Instruction: Sessions Outline and Reference Material, U.S. Government Printing Office, Washington D.C.

Chapter 4

[17] I am grateful to Mr. John Mustafa for contributing this project.

Chapter 5

[18] Monden, Y., *Toyota Production System*, Industrial engineering and Management Press, Atlanta, USA, 1983.

[19] Shingo, S., *Study of Toyota Production System,* Japan Management Association, Tokyo, 1985.

Chapter 6

[20] I am grateful to Mr. Tan Kek Kiong for contributing this project.

[21] Chart adapted from: Wilkins, Dennis; *The Bathtub Curve and Product Failure Behavior*, www.weibul.com, Issue 21, Nov 2002.

Chapter 7

[22] Kano, Noriaki, et al., "Attractive Q Vs Must Be Q". *Hinshitsu (Quality),* Vol.14, No. 2, 1984. Pp 39 - 48.

[23] From a discussion by Tom Peters in his weekly syndicated column, copyright 1987, TPG Communications.

[24] Deming, Edwards, *Out of the Crisis*, Cambridge, Ma, MIT Press, 1982.

[25] Spear, S. and H.K. Bowen, 1999, "Decoding the DNA of the Toyota Production System," *Harvard Business Review*, Sept.-Oct., 77(5), 97-106.

[26] Schneiderman, "Setting Quality Goals", *Quality Progress,* April 1988.

[27] Quote from the author's previous book: *Total Quality Essentials*, McGraw-Hill, New York, 1998.

[28] Spear, S. and H.K. Bowen, 1999, "Decoding the DNA of the Toyota Production System," *Harvard Business Review*, Sept.-Oct., 77(5), 97-106.

[29] *The 10X rule* of thumb is credited to Barry Boehm; it applies to both hardware and software.

[30] Kume, H, Business Management and Quality Cost, the Japanese View, Quality Progress, April, 1985.

[31] Shigeo Shingo, *Study of Toyota Production System*, Japanese Management Association, Tokyo, 1981.

[32] A part of the section on Total Quality is quoted from the author's previous books: *Total Quality Essentials*, McGraw-Hill, New York, 1998; and *TQC-The Asian Experience,* an internal Hewlett-Packard publication.

Chapter 8

[33] John D.C. Little, 1961, *A Proof for the Queuing Formula: $L = \lambda W$,* Operations Research, Vol. 9, No. 3, pp. 383 – 387.

[34] Hopp, W.J., Spearman, M.L., 2001, *Factory Physics*, McGraw-Hill, New York.

[35] John Miltenburg, One-piece flow manufacturing on U-shaped production lines, IEE transactions, 2001, 33, 303-321.

[36] Askin, R, Standridge, C, 1993, *Modeling and Analysis of Manufacturing Systems*, Wiley, New York, USA.

[37] Hopp, W.J., Spearman, M.L., 2001, *Factory Physics*, McGraw-Hill, New York.

Chapter 9

[38] Quote from: The International QC Forum, Nov. 1984.

[39] Quote from New York Times, May 26, 2010: "*Electronics Maker Promises Review after Suicides.* With about 800,000 Chinese employees, revenue of about $60 billion a year and a reputation for military-style efficiency, Foxconn is possibly the world's biggest electronics maker. It is now also the focus of criticism and troubling questions about a wave of suicides among its workers at a pair of factories here that serve as major suppliers to global brands."

Chapter 10

[40] Charles F. Knight, "*Emerson Electric: Consistent Profits, Consistently*". Harvard Business Review, Jan./Feb. 1992, pp. 65-70.

[41] Much of the Hoshin planning methodology mentioned here comes from the author's personal experience and previous book: "*TQC - The Asian Experience*", an internal Hewlett-Packard publication. The categories (objectives, goals, strategies, and performance measures) that we show are similar at companies that use Hoshin plan methodology in the USA and Japan. Their forms, however, may be different – hence, we have also shown alternative Hoshin planning forms and matrix.

[42] *The Itoh model*: We are not able to trace the source of this model; we have seen it many times without reference to its original source.

[43] Quote from a CEO in: Dettmer, W. H., *Strategic Navigation: A Systems Approach to Business Strategy*, ASQ Quality Press, Milwaukee, Wisconsin, 2003.

[44] Dettmer, W. H., *Strategic Navigation: A Systems Approach to Business Strategy*, ASQ Quality Press, Milwaukee, Wisconsin, 2003.

[45] I am grateful to Dr. Dieter Legat, of the Delta Institute Switzerland, for his inputs on TOC and Hoshin Planning. Dr. Legat is a long time practitioner of both Hoshin Planning and TOC. He says that he regrets only two things in this field – (a) that TOC is called a theory rather than a management practice (because that turns off business leaders) and (b) that TOC and Hoshin Planning are positioned as alternatives, rather than mutually amplifying practices.

[46] Deming, Edwards, *Out of the Crisis*, Cambridge, Ma, MIT Press, 1982.

Chapter 11

[47] Womack, James; Jones, Daniel; 1996, *Lean Thinking: Banish Waste and Create Wealth in Your Corporation:* Simon & Schuster, New York, New York. The definition is taken from the text, but we have shortened it.

Chapter 12

[48] Pay, Rick, "Everybody's jumping on the Lean Bandwagon, but many are being taken for a ride", Industry Week, May 1, 2008.

[49] Quote from Robert Miller, Executive Director of the *Shingo Prize for Operational Excellence,* interviewed on radiolean.com, July, 2010.

[50] Hines, Peter; Holweg, Matthias; Rich, Nick; *Learning to Evolve: A Review of Contemporary Lean Thinking;* Production Management, Vol. 24 No. 10, 2004, pp. 994-1011.

Printed in Great Britain
by Amazon.co.uk, Ltd.,
Marston Gate.